Emerging Actuator
Technologies

Emerging Actuator Technologies
A Micromechatronic Approach

José L. Pons

John Wiley & Sons, Ltd

Other Wiley Editorial Offices

John Wiley & Sons Inc., 111 River Street, Hoboken, NJ 07030, USA

Jossey-Bass, 989 Market Street, San Francisco, CA 94103-1741, USA

Wiley-VCH Verlag GmbH, Boschstr. 12, D-69469 Weinheim, Germany

John Wiley & Sons Australia Ltd, 33 Park Road, Milton, Queensland 4064, Australia

John Wiley & Sons (Asia) Pte Ltd, 2 Clementi Loop #02-01, Jin Xing Distripark, Singapore 129809

John Wiley & Sons Canada Ltd, 22 Worcester Road, Etobicoke, Ontario, Canada M9W 1L1

Wiley also publishes its books in a variety of electronic formats. Some content that appears
in print may not be available in electronic books.

Library of Congress Cataloging-in-Publication Data

Pons, José L.
    Emerging actuator technologies: a micromechatronic approach / José L. Pons.
        p. cm.
    Includes bibliographical references and index.
    ISBN 0-470-09197-5 (alk. paper)
    1. Mechatronics. 2. Actuators. I. Title.

    TJ163.12.P66 2005
    621 – dc22

                                                    2004062877

British Library Cataloguing in Publication Data

A catalogue record for this book is available from the British Library

ISBN 0-470-09197-5

Produced from LaTeX files supplied by the authors and processed by Laserwords Private Limited,
Chennai, India
Printed and bound in Great Britain by TJ International Ltd, Padstow, Cornwall
This book is printed on acid-free paper responsibly manufactured from sustainable forestry
in which at least two trees are planted for each one used for paper production.

*To Amparo
and our children*

Contents

Foreword

In recent years, new physicochemical principles and new transducing materials have been discovered, which make it possible to generate mechanical actions that perform the basic functions of an actuator. In today's world, with increasingly stringent demands for control of widely varying devices, there is a need to find ever more efficient actuators, with more power, bandwidth and precision but smaller in size. This is clearly the case, for example, of actuators for implantation in human beings or for use on space vehicles.

The scientific approaches that the research community has adopted toward the new actuators have been very unfocused and sectoral, as readers will appreciate from the long list of over 50 very recent references dealing with specific aspects that are discussed at the end of this book. This generalized situation of fragmented analysis contrasts with the painstakingly comprehensive and rigorous account, which, here, offers the reader an overview of the subject. Its purpose is to help build up a body of doctrine relating to emerging actuator technologies, and its primary virtue is to treat the various different materials as active or semiactive mechatronic devices so as to be able to integrate them in a controlled system. The actuator itself is considered as a mechatronic system with all its attendant derivatives.

As to the content of the book, this deals systematically with all the principal types of advanced actuators. In methodological terms, each chapter analyzes the principles of transduction with reference to their origin, the materials made, the equations and their characteristics; it then deals with the corresponding control circuits and devotes considerable space to details and novel aspects of applications. Of these, we could mention for example piezoelectric elements for ultraprecise (nm) positioning in grinding machine tools, or shape memory actuators (SMA) for automatic oil-level control in high-speed trains, or, again, magnetorheological fluids (MRFs) for use as active shock absorbers in a lower-limb prosthesis to adapt to an amputee's gait. In every case, the author provides details of performance and even references to the makers of the actuators described.

The success of this integrated approach is undoubtedly a result of the considerable experience of the author, a prominent member of the SAM (Sensors, Actuators and Microsystems) research group at the Industrial Automation Institute of Madrid (affiliated to the National Science Research Council, CSIC), who has taken part in and directed numerous projects in this area of research and has worked with and at the most prestigious European and American research organizations and universities. It was his vocation as a researcher that first drew him to so innovative a field and to follow its progress. Conscious of the interest that the theoretical knowledge acquired will attract now and in the future, and also of their practical importance, the author conscientiously explains the most basic ideas clearly and concisely, and moves from there to other increasing complex notions, always highlighting the strengths and weaknesses of these new technologies.

The book commences by presenting the subject of actuators in a general way and explaining their function as a mechanical correcting element in a controlled system. It discusses the dual actuation and sensing functions of certain smart materials, and also the different kinds of actuators, their parameters and the criteria with which they are evaluated.

It then goes on to analyze piezoelectricity as a basis for the development of actuators, both resonant and nonresonant, which react to the application of an electric field; shape memory actuators (SMAs) and the different alloys that possess this ability to actuate when subjected to thermal changes; electro active polymers (EAPs), either ionic or electronic, in which the different effects of the interchange and ordering of matter is especially important; actuators made from electro- and magneto rheological fluids (ERFs, MRFs), whose rheological characteristics vary, depending on the external fields applied; and actuators based on magnetostriction, either positive or negative, where magnetic domains are reoriented by means of an external magnetic field.

Having described their characteristics, the book embarks on an invaluable comparative study of all these actuators, noting the unsolved problems and the latest trends in their resolution. It places particular emphasis on control, on drivers, and, where applicable, on performance or quality standards of actuators. These qualities of the book alone are sufficient to explain its utility to researchers and designers of actuators, or simply to anyone interested in advanced automatic control systems.

If knowledge is the basis of the future, then this book will help us attain that knowledge by furnishing an excellent base from which to embark on new research, development and applications in the vast universe of advanced actuators.

Ramón Ceres
Research Professor, CSIC

Preface

My first contact with the world of new actuators dates back to 1995 when, during a research visit to the Mechanical Department at the Katholieke Universiteit Leuven, Belgium, I was astonished by the elegant and incredibly simple operation of shape memory actuators (they were being applied to biomedical devices). Ever since, I have become acquainted with ever more new types of actuators during research visits to MIT, USA (polymer gel actuators, 1996), TU München, Germany (shape memory actuators, 1997 and 2000), Scuola Superiore di Studi e Perfezionamento Sant'Anna, Italy (micromechatronics of sensing and actuation, 1998 and 1999) and the Department of Cybernetics, University of Reading, UK (magnetorheological actuators, 2002).

These visits and research activities at my home institution made me realize that to be sound an approach to the world of emerging actuator technologies (EATs) must be accompanied by an engineering-based approach that is only realizable if EATs are conceived as true mechatronic systems.

The purpose of this book is to provide an introductory view, with a clear mechatronic focus, of the various different new actuator technologies (piezoelectric, shape memory, electroactive polymer, magnetostrictive and electro- and magnetorheological actuators). It is intended as a reference for mechanical, electrical, electronic and control engineers designing novel actuator systems. The book highlights the concurrent need of all these disciplines for a sound application-oriented approach to the development of new actuators. As such, it covers the principles of actuation, the governing equations, the mechatronic design of actuators and control strategies, their analysis in terms of performance and their behavior upon scaling, and it analyzes the application domains for each technology.

The comprehensive analysis of emergent actuators that this book offers is unique in its scope and in its specific focus on applications, with an unparalleled comparative study of emerging technologies with one another and with traditional actuators.

The book is organized in seven chapters. The first chapter introduces the different concepts and aspects to be considered in the analysis, design, control and application of emerging actuators. Chapters 2 through 5 describe emerging active actuators (piezoelectric, shape memory, electroactive polymer and magnetostrictive actuators respectively) and Chapter 6 describes semiactive emerging actuators (electro- and magnetorheological actuators). Finally, Chapter 7 summarizes the most important features, provides a detailed comparative analysis of emerging actuators and analyzes the most probable trends in terms of research activities and application domains.

The writing of this book would have not been possible without help and contributions from many people. I wish to express my gratitude to D. Reynaerts (KULeuven) for his support and contributions during these years, and in particular during my stay at his laboratory in 2004. Many people contributed to this book, especially in connection with the applications and case studies (F. Claeyssen – Cedrat Technologies; J. Peirs – KULeuven; K.D. Wilson – MRS Bulletin; G.L. Hummel and L.C. Yanyo – Lord Corporation; U. Zipfel – Argillon GmbH; J.N. Mitchel – SRI; K. Otsuka and T. Kakeshita; E. Pagounis – Adaptamat Ltd; W. Harwin – U. Reading; S. Skaarup – DTU; D. Mesonero-Romanos – IAI-CSIC). I wish to thank all of them.

Finally, I would like to thank R. Ceres for encouraging me to write this book and for awakening my interest in research, all my colleagues at IAI-CSIC (in particular, the SAM group), my family and my parents to whom I owe everything and to God.

List of Figures

List of Tables

1

Actuators in motion control systems: mechatronics

Actuators are irreplaceable constituents of mechatronic motion control systems. Moreover, they are true mechatronic systems: that is, concurrent engineering is required to fully exploit their potential as actuators.

This chapter analyzes the actuator as a device included in motion control systems. It introduces the intimate relationship between transducers, sensors and actuators, and discusses the implications of sharing these functions on the same component.

It also discusses the role of the actuator as a device establishing an energy flow between the electrical and the mechanical domain, and it introduces a set of relevant performance criteria as a means for analyzing the performance of actuators. These criteria include both static and dynamic considerations, and also the performance of the actuator technology upon scaling.

Actuators are classified into active and semiactive actuators according to the direction in which energy flows through the actuator. Active technologies (Piezoelectric, SMA, EAP and magnetostrictive actuators) are then discussed in Chapters 2 through 5, and semiactive technologies (ER and MR actuators) in Chapter 6.

Finally, after explaining the distinction between emerging and traditional actuators, this chapter concludes with an analysis of other actuator technologies (electrostatic, thermal and magnetic shape memory actuators) not specifically dealt with in separate chapters.

Emerging Actuator Technologies: A Micromechatronic Approach   J. L. Pons
© 2005 John Wiley & Sons, Ltd

1.1   What is an actuator?

The mechanical state of a system can be defined in terms of the energy level it has at a given moment. One possible way of altering the mechanical state of a system is through an effective exchange of energy with its surroundings. This exchange of energy can be accomplished either by passive mechanisms, for example, the typical decaying energy mechanism through friction, or by active interaction with other systems. An actuator is a device that modifies the mechanical state of a system to which it is coupled.

Actuators convert some form of input energy (typically electrical energy) into mechanical energy. The final goal of this exchange of energy may be either to effectively dissipate the net mechanical energy of the system, for example, like a decaying passive frictional mechanism, or to increase the energy level of the system.

An actuator can be seen as a system that establishes a flow of energy between an input (electrical) port and an output (mechanical) port. The actuator is transducing some sort of input power into mechanical power. The power exchange both at the input and output ports will be completely defined by two conjugate variables, namely, an effort (force, torque, voltage etc.) and a flow (velocity, angular rate, current, etc.). Eventually, some input power will be dissipated into heat. See Figure 1.1 for a schematic representation of the actuator.

The ratio of the flow to the effort (conjugate variables) is referred to as *impedance*. If an electrical input port is considered, the voltage and the current drawn will completely define the power flowing in the actuator, and the ratio is the familiar electrical impedance. By analogy to the electrical case, at the mechanical port, the ratio of flow (velocity or angular rate) to effort (force or torque) is referred to as *mechanical impedance*, and both variables will define the power coming out of the actuator.

The concept of power exchange at the input and output ports of an actuator gives rise to a wider definition of actuators as devices whose input and output ports exhibit different impedances. In general, neither the input electrical impedance of an actuator will match that of the controller nor the output mechanical impedance will match that of the driven plant. This lack of match between input and output

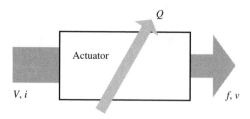

Figure 1.1   Actuator concept: energy flows from the input to the output port. Eventually, some energy is dissipated (undesired losses).

impedances means that impedance adaptation is required both at the input and output ports. The issue of impedance matching will be dealt with in more detail in Section 1.3.

Actuators are most often found in *motion control systems, (MCS)*. In these systems, the ultimate objective is to drive the plant along some reference trajectory. The role of the actuator in such a system is to establish the flow of power by means of some control actions (inputs) in response to process models or sensory data so that the desired trajectory is effectively accomplished.

The dynamic interaction between the actuators and the controlled system can be defined according to the magnitude of the energy being exchanged, $dW = F \cdot dX$, Hogan (1985a). In some particular situations, the instantaneous energy exchange can be ignored. This is the case where the force or the displacement is negligible. On the one hand, if the force is zero ($F = 0$), the system can be considered position controlled. On the other hand, wherever the displacement is zero ($dX = 0$), the system is force controlled.

In general, the interaction will take place with a finite, nonzero instantaneous energy exchange ($dW \neq 0$). In such a case, the motion control system will be able to impose the effort (force, torque), the flow (velocity, angular rate), or the ratio between them (the impedance), but not both simultaneously.

Depending on the sign of the admissible instantaneous work exchange, dW, actuators can be classified as:

1. *Semiactive actuators*: the work exchange can only be negative, $dW \leq 0$. In practice, this means that semiactive actuators can only dissipate energy as a consequence of mechanical interaction with the controlled system. These actuators are dealt with in Chapter 6.

2. *Active actuators*: the work exchange can take any positive or negative value, $dW \lessgtr 0$. For practical purposes, this means that active actuators can either increase or decrease the energy level of the controlled system.

The ultimate constituent of an actuator is the transducer. A transducer has been defined (Middlehoek and Hoogerwerf (1985)) as *a device, which transforms nonelectrical energy into electrical energy and vice versa*. This definition of a transducer emphasizes the fact that most actuators (transducers) are driven by logic elements in which the information flow is electronically established. As such, transducers ultimately transform to and from electrical energy.

Transducers have also been defined (Rosenberg and Karnopp (1983)) as *devices, which transform energy from one domain into another*. Rosenberg's definition of a transducer is broader than the previous one since it does not restrict transduction to or from the electrical domain. Finally, the broadest definition of a transducer makes a distinction between different types of energy within a single domain (differentiating between rotational and translational mechanical energy). It states (Busch-Vishniac (1998)) that a transducer is *a device, which transforms energy from one type to another, even if both energy types are in the same domain*.

A transducer can be used to monitor the status of a parameter in a system, or it can be used to define the status of such a parameter. It is the former use of transducers that produces the *concept of sensors*. A sensor is thus a transducer, which is able to monitor the status of a system (ideally) without influencing it.

On the other hand, an actuator can be defined as a transducing device, which is able to impose a system status (ideally) without being influenced by the load imposed on it.

The transduction process can be established between any two energy domains (see second definition of transducers) or even between different energy types within the same domain (see third definition). Whenever a transducer is used to impose a status on a system (actuator concept), such wide definitions of transducers would include actuators capable of establishing energy flow between any two energy domains. Throughout this book, the first definition of transducers is used and is restricted to output mechanical energy.

A transducer might establish energy flow between nonelectrical input energy domains and output mechanical energy – see the case of a thermally actuated shape memory alloy (SMA) transducer. For practical purposes, it is always possible to include any subsystem in charge of electrically driving the transducer in the actuator system. This applies, for instance, to electrically heating the SMA transducer by means of a Joule effect (delivering heat through resistance heating). The actuator system as a whole establishes a flow of energy between the electrical domain and the mechanical domain (see Figure 1.2).

The use of electrical energy at the input port of actuators has clear advantages:

1. *Compatible energy domains.* Most motion control systems (in which actuators are usually included) are controlled electronically; thus, the output energy domain of the control part is already in the same energy domain as the input actuator port.

2. *Fast operation of electric devices.* Electronic and electric devices are characterized by fast operation, in most cases much faster than the intrinsic time constants of the actuator. This improves the controllability of actuators.

3. *Availability of components.* The electronic components used in the control and conditioning system are well-known and readily available.

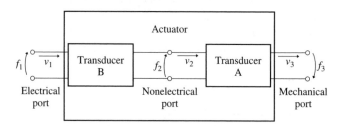

Figure 1.2   Actuator concept as a two-port transducer: input electrical port and output mechanical port.

In view of the above considerations, the actuator concept that we will use throughout this book comprises both the transducer itself (possibly between non-electrical domains and the mechanical domain) and the subsystems responsible for electrically driving the transducer. Note that the subsystems used for electrically driving the transducer could, in turn, be considered an additional transducer, according to the broad definitions noted earlier (see transducer B in Figure 1.2).

1.2 Transducing materials as a basis for actuator design

Transduction is the process of energy conversion between either different energy domains (for instance, from thermal to mechanical energy) or different energy types within the same domain (for instance, between rotational and translational energy). A more restrictive definition of transduction defines the input domain as electrical energy and the output domain as mechanical energy.

The transducer is the device in which transduction is accomplished. In general, two types of transducers can be considered (Busch-Vishniac (1998)):

1. *Geometrical transducers.* In geometrical transducers, the coupling between input electrical energy and output mechanical energy is based on the exploitation of some geometrical characteristics. Actuators resulting from geometrical transducers are called by extension *geometrical actuators*. This applies to all rotational actuators.

   In particular, if a rotational permanent magnet electromagnetic DC motor is considered, the geometry of the magnetic flux with regard to the configuration of the current flowing in the coils leads to a Lorentz interaction, which, in turn, results in a rotational motion of the coil (see Figure 1.3).

2. *Transducing materials.* A transducing phenomenon between any of the different energy domains is directly exploited to develop actuators. Examples include stack piezoelectric actuators or shape memory alloy actuators directly pulling a load.

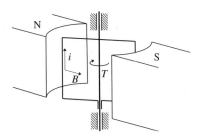

Figure 1.3   A DC motor as a geometrical transducer.

In most cases, the time lapse between the discovery of a new transducing material and its eventual application in the development of actuators might be decades or even centuries. This is the case, for instance, of ER and MR fluid actuators (see Chapter 6). The modification of the rheological behavior of both types of fluid in response to electric or magnetic fields, respectively, has been known since the late 1940s. Only recently have they been applied industrially in the context of vibration isolation (see Case Study 6.3, page 240).

1.2.1   Energy domains and transduction phenomena

Most often when dealing with transducers, seven main energy domains are considered, namely, chemical, electrical, magnetic, mechanical, optical, fluid and thermal. Transduction can be found between any two of these energy domains. In addition, different transduction phenomena are possible for a given pair of energy domains. Some of the transduction processes of interest when developing actuators are briefly analyzed in the following paragraphs.

1. *Thermomechanical transduction.* In this energy conversion process, the input energy is in the thermal domain and the output energy in the mechanical domain. Several actuators can be developed by following this conversion scheme:

   (a) *Shape memory alloy (SMA) actuators.* In this type of actuators, the input thermal energy triggers a phase transition in the alloy, which results in the shape recovery of a previously deformed state. These actuators are discussed in detail in Chapter 3.

   (b) *Thermal actuators.* In this type of actuators, the different thermal expansion coefficients of two thin metallic laminas cause a bending of the composite structure upon heating and cooling. These actuators are described in more detail in Section 1.10.2.

   (c) *Thermally active polymer gels.* Some polymer gel actuators respond to thermal stimuli. These are reviewed in more detail in Chapter 4.

   (d) *Thermal expansion actuators.* It is well-known that temperature changes cause expansion–contraction of all materials. Thermal expansion can be considered a direct thermomechanical transduction process.

2. *Magnetomechanical transduction.* These actuators establish an energy flow from the magnetic domain to the mechanical domain and vice versa. Again, several actuators can be developed, depending on various different transduction phenomena:

   (a) *Magnetostrictive actuators.* Magnetostrictive actuators exhibit a reorientation of magnetic dipoles in the presence of an externally imposed magnetic field. Magnetic domain reorientation results in extension-contraction in the dominant direction. These actuators are analyzed in Chapter 5.

(b) *Magnetorheological fluid (MRF) actuators.* MRFs exhibit changes in their rheological properties when subjected to external magnetic fields. The apparent viscosity of these materials is thus modified according to the magnetic field. They are semiactive actuators: that is, they can only dissipate energy. MRF actuators are discussed in Chapter 6.

(c) *Magnetic shape memory alloy (MSMA) actuators.* In most instances, MSMAs are considered a subclass of magnetostrictive actuators. However, they exhibit very different actuator characteristics and are evolving into an independent new class of actuators. They are addressed in Chapter 1.

3. *Electromechanical transduction.* The energy in the input electrical domain is transformed into mechanical energy. In most of the following actuator technologies, the transduction process is reversible. Some of the technologies listed below are used concomitantly with the converse transduction process in what are known as smart actuators (see Section 1.3).

(a) *Electromagnetic actuators.* The Lorentz interaction between a flowing electrical charge and a magnetic field is exploited to supply either translational or rotational mechanical energy to the coil. The magnetic field can be established either by means of permanent magnets or by a second coil. This is a well-known, traditional actuator technology, and in this book, it is only mentioned as a reference for comparison with emerging technologies.

(b) *Piezoelectric actuators.* The converse piezoelectric effect resulting from the interaction of an imposed electric field and electrical dipoles in a material results in a deformation. This deformation is used to drive the plant. The converse piezoelectric effect can be used directly or through geometrical transducer concepts. This is analyzed in detail in Chapter 2.

(c) *Shape memory alloy (SMA) actuators.* These actuators have already been mentioned in connection with thermomechanical transduction. Thermal energy is usually supplied through resistive heating (Joule effect), and, hence, these can also be considered electromechanical transducers.

In the context of smart actuators, a linear relationship between the electrical resistance and the displacement is used to establish a sensor model.

(d) *Electroactive polymer (EAP) actuators.* Within the broad family of EAP actuators, dry type polymers directly exploit Maxwell forces or the electrostrictive phenomenon to obtain mechanical energy from electrical input energy. In addition, some ionic EAPs (also referred to as wet EAPs) are triggered by small electric fields. All these actuators are presented in Chapter 4.

(e) *Electrorheological fluid (ERF) actuators.* Like MRF actuators, the rheological properties of ERF actuators are altered when an electric field is applied. Again, these are semiactive actuators and so can only dissipate the energy of the plant. They are analyzed in Chapter 6.

4. *Fluid-mechanical transduction.* Some traditional actuators (pneumatic and hydraulic actuators) convert the pressure of a fluid into mechanical energy, either rotational or translational.

1.2.2   Transducer basics

A transducer is a two-port device. Unless it is intended to transduce between different types of energy within the same domain, it will have four terminals.

As noted earlier, two conjugate variables (effort, v, and flow, f) at each port will define the power entering and leaving the transducer. A general scheme of this two-port device is shown in Figure 1.4.

For a linear transducer, the relationship between effort and flow at both terminals will have the following form:

$$\begin{bmatrix} v_1 \\ f_1 \end{bmatrix} = \begin{bmatrix} t_{11} & t_{12} \\ t_{21} & t_{22} \end{bmatrix} \begin{bmatrix} v_2 \\ f_2 \end{bmatrix} \tag{1.1}$$

According to the structure of the two-dimensional matrix defining a particular transducer, these can be classified into the following:

- *Transforming Transducers.* In a transforming transducer, the following relations between coefficients in the transducer matrix are met:

$$t_{12} = t_{21} = 0 \quad \text{and} \quad t_{22} = -\frac{1}{t_{11}} \tag{1.2}$$

From the relationships of Equation 1.2, it follows that the flow in port 1, v_1, is linearly related to the flow in port 2, v_2, k being the constant of

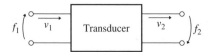

Figure 1.4   Two-port transducer: input power defined by conjugate variables f_1 and v_1 and output port power defined by variables f_2 and v_2.

proportionality. Likewise, the effort in port 1 is related to the effort in port 2 by a constant, which is the negative reciprocal of k:

$$\begin{bmatrix} v_1 \\ f_1 \end{bmatrix} = \begin{bmatrix} k & 0 \\ 0 & -1/k \end{bmatrix} \begin{bmatrix} v_2 \\ f_2 \end{bmatrix} \tag{1.3}$$

- *Gyrating Transducers.* In the case of gyrating transducers, the following relations apply:

$$t_{11} = t_{22} = 0 \quad \text{and} \quad t_{21} = -\frac{1}{t_{12}} \tag{1.4}$$

In gyrating transducers, the flow in port 1 is linearly related to the effort in port 2 by a constant, g. Likewise, the effort in port 1 is linearly related to the flow in port 2, the constant of proportionality being the negative reciprocal of g. The following relationship between input and output conjugate variables is found in gyrating transducers:

$$\begin{bmatrix} v_1 \\ f_1 \end{bmatrix} = \begin{bmatrix} 0 & g \\ -1/g & 0 \end{bmatrix} \begin{bmatrix} v_2 \\ f_2 \end{bmatrix} \tag{1.5}$$

Equations 1.3 and 1.5 impose a causality condition between the conjugate variables of input and output ports. In particular, for the transforming transducer they indicate that only one of the flow variables v_1 or v_2 can be chosen as independent variables. Once a flow is specified as an independent variable, the flow in the other port follows from the causal relation of Equation 1.3. The same can be said for the efforts: that is, they are causally determined by Equation 1.3:

$$v_1 = k \cdot v_2 \tag{1.6}$$

$$f_1 = -1/k \cdot f_2 \tag{1.7}$$

For gyrating transducers, the causality relations of Equation 1.5 impose conditions between flow and effort in both ports. If a flow is chosen as an independent variable, the effort at the other port is causally determined:

$$v_1 = g \cdot f_2 \tag{1.8}$$

$$f_1 = -1/g \cdot v_2 \tag{1.9}$$

The results of transforming and gyrating transducers can be exploited to develop concomitant sensing and actuation in some instances. This is explained in more detail in the next section.

Different conjugate variables are defined according to the particular input and output energy domains of a particular transducer. The flow and effort corresponding to the most common energy domains are summarized in Table 1.1. Note that the definition of conjugate variables as those defining the power at each port does not apply to the case of the thermal energy domain.

Table 1.1   Conjugate variables in transducer ports for various energy domains.

Energy domain	Flow, v	Effort, f
Electrical	Current, i	Voltage, V
Fluid	Volume flow rate, Q	Pressure drop, P
Magnetic	Magnetic flux, Φ	Magnetomotive force, $\mathcal{F}$
Mechanical translational	Velocity, v	Force, F
Mechanical rotational	Angular velocity, ω	Torque, T
Thermal	Heat flow, q	Temperature, T

A complementary method of analyzing actuators is by means of the electric-circuit analogy. In the electric-circuit analogy, ideal effort sources (voltage sources in the electrical terminal and force sources in the mechanical counterpart) are used to drive an electrical and a mechanical impedance, Z_e and Z_m respectively. The result is a current flowing in the electrical port, I, and a mechanical flow, that is, velocity, v, in the mechanical port.

In the electric-circuit analogy, both ports are coupled through a black box representing the transduction process (see Figure 1.5). Thus, in the electric-circuit analogy, the equations describing the relationship between flow and effort in both the electrical and the mechanical domain are

$$V = Z_e I + T_{em} v \qquad (1.10)$$

$$F = T_{me} I + Z_m v \qquad (1.11)$$

In Equation 1.10, V is the voltage applied to the electrical terminal to drive the actuator, Z_e is the so-called blocked electrical impedance of the actuator, I is the electrical current flowing, T_{em} is the transducing coefficient from the mechanical to the electrical domain, F is the force acting on the actuator's mechanical port, T_{me} is the coefficient of transduction from the electrical to the mechanical domain, Z_m is the mechanical impedance and, finally, v is the velocity at the mechanical port.

The electrical driving-point impedance, Z_{ee}, is defined (Hunt (1982)) as the ratio of the applied voltage to the current drawn when all the electromotive terms in the transduction process are suppressed. The mathematical formulation for the

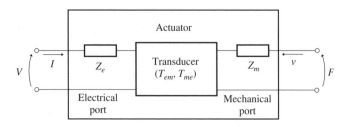

Figure 1.5   Electric-circuit representation of actuators.

driving-point electrical impedance is

$$Z_{ee} = \left(\frac{V}{I}\right)_{F=0} = Z_e + \frac{-T_{em}T_{me}}{Z_m} \qquad (1.12)$$

The first term in the driving-point electrical impedance is the *blocked electrical impedance*, Z_e, which is defined as the electrical impedance of the actuator when the mechanical displacement is set to zero. The second term in the driving-point electrical impedance depends on three factors: the two transduction coefficients, T_{em} and T_{me}, and the mechanical impedance, Z_m.

The second term in the driving-point impedance is what is called the *motional impedance*, Z_{mot}. The fact that the motional impedance depends on the mechanical impedance, Z_m, indicates that even in the absence of external forces the actuator's electrical impedance will be affected by the motion of the mechanical port (Pratt (1993)). This dependence on the motion is, however, expected to be low where the actuator is driven at low frequencies.

The analysis of actuators in terms of the electric-circuit analogy is of particular interest for purposes of developing the *smart actuator* concept (see Section 1.5). In this context, a model of the actuator's blocked impedance can be used to estimate the position (or velocity) at the mechanical port, thus allowing the actuator to be used concomitantly as a sensor.

1.3   The role of the actuator in a control system: sensing, processing and acting

Motion control systems can be regarded as the paradigm in the application of actuators. In Systems Theory, the element whose motion is to be controlled is denoted as the *plant*. The ultimate objective of a motion control system is to drive the plant according to a desired reference *trajectory*.

Here, the term *trajectory* must be considered in the widest sense. As introduced in the previous section, the control system will only be capable of imposing one of the two conjugate variables defining the mechanical interaction with the plant at the actuator output port. Depending on how this interaction is established, the *trajectory* might be a collection of successive positions (flow) of the plant parameterized by the time. It could also be a reference force (effort) parameterized by time, and it may occasionally be some sort of relation (impedance) between the flow and the effort as a function of time.

In general, motion control systems drive the plant according to the reference trajectory by means of a combination of functions: sensing, processing and actuation. Some of these functions may not be present in a particular motion control system. This is true of the *feed-forward control* of a plant. In feed-forward control schemes, detailed and accurate models of the plant and the actuator are available so that it is possible to predict the driving signal needed to attain the reference trajectory. Sensors are therefore not strictly necessary.

The role of the different components in MCS is discussed in the following paragraphs:

1.3.1 Sensing

As noted previously, sensors are transducing devices that monitor the status of a parameter of the plant. In a general motion control scheme, the reference trajectory must be compared to the actual one. Reactive measures to counteract deviations can then be implemented on the basis of this comparison.

The use of sensors in *feedback control* motion systems provide means for improving the robustness of the whole process. Additionally, they enable the implementation of disturbance rejection strategies. Feed-forward control schemes (sensorless) are susceptible of being affected both by inaccuracy in the plant and actuator models and by external (to the plant and the actuator) and internal disturbances. On the contrary, feedback schemes are much more robust against disturbances and can make allowance for inaccuracy in the models.

Sensors monitor the status of the plant, ideally without influencing it. Since they are transducing devices, there will always be a flow of energy through them (a sensor in which no flow of energy is established in either direction is not physically practicable). Since they must operate without affecting the plant, the flow of energy must be minimized. Consequently, they are mostly low-power, miniaturized devices.

1.3.2 Processing

The processing function in a motion control system is done by the controller. The controller provides the equivalent to the intelligence in a control system. It usually receives the reference trajectory as an input (most likely from an upper-level task planner in a hierarchical control approach) and computes the required action to drive the plant according to the reference trajectory.

The controller in feedback schemes obtains information on the status of the plant through sensors. On the basis of this information, the deviation from the reference trajectory is calculated and corrective actions implemented. Corrective actions take the form of input energy in the domain of the actuator's input port.

The controller in feed-forward schemes computes the driving actions according to the reference trajectory and models establishing the relationship between control actions and plant trajectory. The controller in this scheme might receive sensor data but is not used in a reactive approach to counteract possible deviations from the reference trajectory.

In controlling the flow of energy to or from the plant, the controller modulates the input power to the actuator. Since the input power at the electrical port will be determined by both the flow (current) and the effort (voltage), one possible approach is to command the level of input current or voltage.

The choice of the input conjugated variable to be controlled has direct implications on the performance of the actuator and, hence, on the plant. A typical example of this situation can be found in current- or voltage-controlled piezoelectric actuators (see Chapter 2). In piezoelectric actuators, voltage control produces hysteretic behavior, while current or charge control produces linearized plants.

In practice, electrical power supplies with variable voltage are not readily available. An alternative means of modulating the input power is to switch between discrete levels and produce averaged values. Moreover, in some instances, switching techniques are the only way to effectively drive an actuator (see Chapter 4). The rationale of using switching techniques is described in more detail in Section 1.4.

1.3.3 Actuation

In a motion control system, the actuation function is accomplished by the actuator. The actuator is the only unavoidable component of a motion control system. As previously defined, the actuator establishes a flow of energy between the electrical and the mechanical domains. The function of the actuator is to impose a state on the plant, ideally without being affected by the load.

The plant is driven according to the reference trajectory, by either increasing or decreasing the energy level of the plant. The particular way this energy flow is established is determined by the joint interaction of the plant and the actuator at the output port.

This can be easily illustrated by the simple example of an actuator driving a mass. The actuator can impose an effort (force) on the mass. However, the flow (velocity) of the plant will be determined by the inertial characteristics of the mass. Here, the mass (plant) is acting as an admittance (it receives an effort and determines the flow).

In most situations, the plant acts as an admittance. According to the *principle of causality* (Hogan (1985a)), *the action of the motion control system on the plant must be complementary to it*, that is, wherever the plant is an admittance, the control system behaves as an impedance (accepts a flow and imposes an effort).

The actuator ideally will not be affected by the load. In practice, the load will impose limits on the actuator's power delivery. Because of its dependence on the load, the actuator usually behaves as a low pass filter: that is, in the frequency domain, actuation is only possible up to some cutoff frequency, which is closely related to the device's mechanical time constant (see Figure 1.6). The dependence on the load can also be seen in the typical effort (force) versus flow (velocity) curves of all actuators. Figure 1.7a shows the torque–velocity curve for a permanent magnet DC electromagnetic motor, while Figure 1.7b shows the same curve for a Travelling Wave Rotational Ultrasonic Motor. The available torque is reduced as the actuation velocity is increased, while, ideally, this should be kept constant for all velocities. There is always an effort (stall torque) above which no flow (velocity) is available.

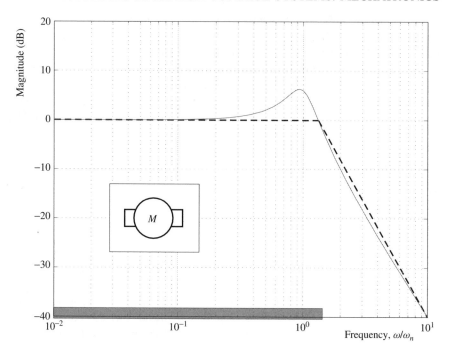

Figure 1.6   Limit in the frequency bandwidth of motors as a consequence of finite power: allowable frequency range highlighted in thick black.

1.3.4   Impedance matching

According to the discussion above, the entire motion control system can be seen as an intelligent modulation of energy flow between an electrical source and the plant, so that it is driven according to a reference trajectory. For this to be properly accomplished, energy flow must be smoothly established. Energy must flow from the electrical power source to the actuator through the input port, and then after being transduced, from the actuator output port to the plant.

Let us recall here the case of a vibrating string fixed at both ends (see Figure 1.8a). When the equilibrium state is altered by imposing a displacement on the string (playing the guitar), a wave is generated and it travels towards both ends (see Figure 1.8b). Since the mechanical impedance of the string and the frame (which holds the string at both ends) are dissimilar, the wave is reflected and a traveling wave in the opposite direction is established. The superposition of traveling waves in opposite directions leads to a standing wave that dissipates all the input energy through frictional mechanisms (see Figure 1.8c).

In this example, the input energy (the energy of the vibrating string) cannot be transferred to the frame (because of the impedance mismatch). The energy is reflected at the interface and is completely dissipated in the string.

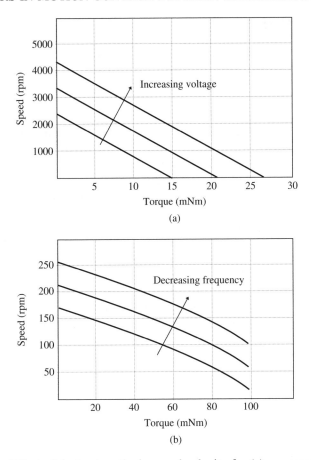

Figure 1.7   Effect of the load on the imposed velocity for (a) a permanent magnet DC motor and (b) an ultrasonic motor.

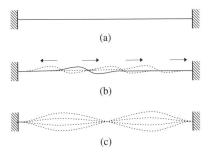

Figure 1.8   Effect of an impedance mismatch on power transmission: the perturbation of the equilibrium in a string (a), results in traveling waves towards the frame (b). Energy is reflected because of the impedance mismatch and dissipated in the string (c).

In controlling the energy flow in motion control systems, an analogy can be established with the previous example. In the mechanical port, the impedance of the plant (ratio of required velocity to force) will, in general, be different from the actuators' output impedance. If no means for matching these impedances is provided, the energy, rather than flowing, will be dissipated at the actuator.

This is exemplified by a permanent magnet DC electromagnetic motor driving a mass (plant) (see Figure 1.9). The typical operating range of DC motors is that of low torque and high velocity, that is, a high mechanical impedance. In general, the driving requirements of plants are low velocity and high torque, that is, low mechanical impedance. If no matching is provided, the DC motor will not be able to impart any motion to the mass, that is, the stall torque will be lower than the required driving torque. All the energy input is then dissipated to heat the motors' armature (see Figure 1.9a). After matching the mechanical impedance at the output port, energy flows and the DC motor can impart a movement to the mass (see Figure 1.9b).

There is a similar situation at the electrical input port. See Chapter 2 for a detailed description of electrical impedance matching when driving piezoelectric actuators in resonance.

Figure 1.10 represents the different constitutive parts of a motion control system for a smooth energy flow from the electrical power source to the plant. Note that

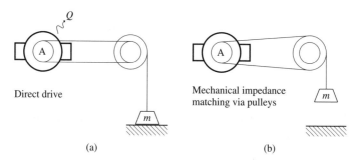

(a)   (b)

Figure 1.9   Mechanical impedance matching: (a) the actuator heats up as power is not transferred to the load and (b) the power can be transmitted to the load after pulley impedance matching.

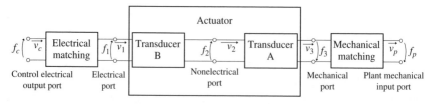

Figure 1.10   Schematic representation of electrical and mechanical impedance matching of an actuator.

the actuator will comprise any requisite subsystem with the function of electrically driving a possible nonelectrical input port transducer.

1.4 What is mechatronics? Principles and biomimesis

The term mechatronics was coined in Japan in the mid-1970s to denote the engineering discipline dealing with the study, analysis, design and implementation of hybrid systems comprising mechanical, electrical and control (intelligence) components or subsystems. Ever since, mechatronics has been understood to mean the integrated and concurrent approach of engineering disciplines for the study of such hybrid systems.

1.4.1 Principles

The black box convention for system analysis is used throughout this book. A system, then, is conceived as a black box with an input and an output port. Systems can interact with one another to set up systems with added complexity and functionality. A mechatronic system is thus a hybrid system including sensor subsystems, control subsystems and actuator subsystems.

Subsystems can be connected in a cascade configuration. In this configuration, there is always an interaction between the output port of one subsystem and the input port of the consecutive one. As discussed earlier, the state of the interaction between two cascaded systems is defined according to conjugated variables.

Mechatronic systems are in many instances synonymous of motion control systems. As such, they include different functional subsystems as introduced in Section 1.3. A very interesting feature of mechatronic systems is the *combination of functions in the same component*. Here, we are especially interested in the possible combination of sensing and actuation functions in a subsystem. This aspect is analyzed in more detail in Section 1.5, when dealing with concomitant sensing and actuation.

The combination of functions by means of mechatronic integration of disciplines in the design of actuators has clear functional benefits. The *miniaturization of systems* can be seen as a direct consequence of this combination of functions. Yet, the paradox of opposite rationales in the design of sensors and actuators must be addressed, see Section 1.5.

This book focuses particularly on analysis of the actuator subsystem within a mechatronic system. The actuator subsystem itself can be shown to exhibit the same mechatronic characteristics as a motion control system. This means that the actuator can be analyzed as a mechatronic system, and it will benefit from the intrinsic cross-fertilization between engineering disciplines (Reynaerts *et al.* (1998)).

The concept of an actuator as a true mechatronic system will be illustrated with the example of a resonant piezoelectric drive. A piezoelectric actuator is an electromechanical device in which the converse piezoelectric effect is used to transduce from electrical to mechanical energy domains (see Chapter 2).

A piezoelectric ceramic is characterized electrically by a capacitive load that is out of resonance and a resistive electrical load that is at resonance (local minimum in mechanical impedance) and antiresonance (local maximum in electrical impedance). Piezoelectric resonators are driven close to their resonance or antiresonance frequencies. In such a driving condition, the electrical load is resistive, and so input voltage and current will exhibit a phase lag close to zero.

In actuators of this type, an applied external load or temperature change will lead to a shift in the resonance frequency (see Figure 1.11). If this is not compensated for, the operating point of the piezoelectric drive will generally not be perfectly tuned. Note the relative position of the original operating point (light grey dot at f_1 in Figure 1.11) with respect to the resonance as compared to the new relative position (grey dot at f_1).

With a mechatronic approach, a self-tuning electrical driver can be designed, which will track any possible fluctuation in the resonance characteristics of the actuator, and, thus, the new operating point will be tuned to the new resonance curve (see grey dot at f_2 in Figure 1.11). In so doing, the phase between voltage and current can be used as an indicator of the electrical impedance of the actuator. This can then be used to close the loop, for instance, by means of a phase locked loop (PLL).

The resonant piezoelectric actuator as described above includes an actuator system (the voltage-driven piezoelectric resonator), a sensor system (monitoring

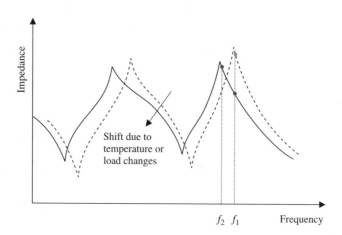

Figure 1.11   Effect of temperature or load fluctuations on the resonance characteristics of piezoelectric actuators and corresponding modification of the operating point.

the phase lag or impedance condition at the input port) and a disturbance rejection control system (the PLL drive).

1.4.2 Mechatronics and biomimesis

Engineering disciplines have always looked to nature as a source of inspiration. Several million years of evolution have seen living creatures progress to their current state. Engineering has very often taken nature as a model and has mimicked biological structures. By way of example, Figure 1.12 shows a biological structure usually taken as a model in the development of helicopter blades and airplane wings.

Mechatronics as an engineering discipline may also benefit from seeking a source of inspiration in nature. As noted earlier, mechatronic systems are in most cases equivalent to motion control systems. As such, the motor control structure of upper mammalians is a perfect model in which to find inspiration.

Hierarchical motor control in mammalians as a model

Hierarchical control schemes are common in motion control systems. There is sufficient evidence to support the view that the structure of the motor control system in mammalians is hierarchically organized. This follows from the excellent performance of both human and nonhuman primates in manipulation tasks. This performance in manipulation comprises, among other functions, superb response to disturbance, for example, increased prehensile force following slippage of grasped objects, and perfect modulation of upper limb (impedance controlled) interaction with the environment.

In these particular manipulative tasks, feedback control schemes involving structures in the central nervous system (CNS) do not seem feasible. In fact, the shortest loop delay involving neural transmission from skin receptors to the CNS and back is in the region of 100–150 ms. If these feedback loops involve computing at the brain level (for instance, in visual feedback operations) the loop delay can reach up to 200–250 ms. With loop delays of this magnitude, the effectiveness

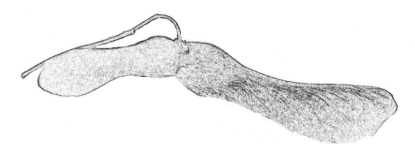

Figure 1.12 Biological model of helicopter blades.

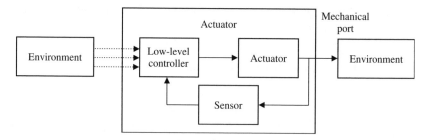

Figure 1.13   Hierarchical representation of the human motor control system.

of feedback modulation of impedance or the response to disturbance would be very much compromised.

Hierarchical control schemes mimicking the structure of the human motor control system are a common approach. Figure 1.13 shows a schematic representation of such a hierarchical control scheme. An upper-level task planner is in charge of sending motion commands (reference trajectories) to low-level controllers. The low-level controllers interact with the plant through sensors and actuators (including the corresponding impedance-matching stages). As seen from the upper-level controller, the process is an open loop.

Switching control of muscle contraction as a model to modulate the input power in actuators

The human musculoskeletal system is driven through switched techniques. Motor stimuli reach the various different muscles through motoneurons. Each muscular stimulus leads in the first instance to muscle contraction followed by relaxation. The time constants of the contraction and relaxation processes are very different. The muscle contraction time constant is much lower than the relaxation time constant; as a result, the musculoskeletal system responses during contraction and relaxation exhibit different dynamics (see the different dynamics in a muscle twitch inset in Figure 1.14).

If repeated stimuli reach the muscle prior to total relaxation, summation occurs, and the result is increasing contraction. Overall muscle contraction is a combination of increased contraction of individual fibers due to summation and increased recruiting of additional motoneurons, and, consequently, muscle fibers. See Figure 1.14 for a schematic representation of the switched control of muscle contraction.

In addition to position control of the human musculoskeletal system, there is sufficient evidence to believe that the modulation of the motor activity in antagonistic muscles is one of the mechanisms that mammalians use to modulate the impedance around an equilibrium position (Hogan (1985b)).

Switching techniques as an approach to modulation of the flow of energy in actuators is of particular interest when actuators exhibit different dynamics in both directions (like in the case of muscle contraction–relaxation). This is generally true

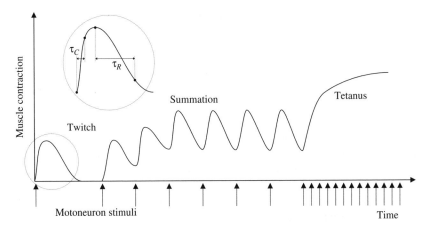

Figure 1.14 Biological model-switching techniques to modulate the flow of energy in actuators.

of thermal actuators, and, in particular, of SMA actuators. In these systems, the time constant for the heating process is generally much lower than for the cooling process (which is limited by thermal inertia and heat dissipation).

Wherever a discontinuity occurs in the change in length of an actuator (see Chapter 4), switching techniques may be the only possible solution for accurate positioning tasks. The discontinuity, as depicted in Figure 1.15, leads to mechanical states that are not attainable in equilibrium. In such a case, switching techniques can maintain the mechanical state without equilibrium within the margin of mechanical state error allowed by the application (Mitwalli (1998)).

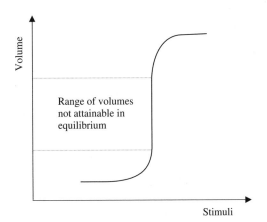

Figure 1.15 Discontinuity in volume–phase transformations leading to nonattainable mechanical states.

Actuation modes based on biological models

The previous two examples of biomimesis are more closely related to the way actuators can be driven and how the energy flow is modulated. Nature is full of models for establishing actuation principles. Here, we will briefly describe two locomotion models that inspired the development of the so-called *inchworm actuators* and *travelling wave linear and rotational ultrasonic motors* (see Chapter 2).

The first model is the locomotion process of some earth worms as depicted in Figure 1.16. This locomotion process is split into two cycles, in one of which the rear and front legs of the worm are fixed alternately to the terrain. In the second cycle, the intermediate segments of the worm elongate and contract alternately. Both cycles are nested to provide the locomotion.

The same principle is followed in the development of inchworm piezoelectric motors (see Section 2.4.3). Here, three independent piezoelectric ceramics are used to mimic the operation of rear and front legs (ceramics 1 and 3), and the intermediate segments (ceramic 2). The piezoelectric actuators 1 and 3 are driven according to the first cycle so that they clamp the rotor (displacer) alternately. The piezoelectric actuator 2 is driven according to the second cycle, mimicking the elongation and contraction of the intermediate segments of the worm.

The second locomotion principle is found in some millipedes and centipedes (see Figure 1.17a). The motion of the different legs is coordinated to produce an approximate sinusoidal pattern in both the elevation and the forward–backward movement. These sinusoidal movements in perpendicular directions produce an elliptic movement of each leg. This elliptic movement (which is implemented in successive legs with a small delay) provides incremental traction to the millipede.

The same principle is exploited in travelling wave ultrasonic motors. The case of linear piezoelectric motors based on this principle is described in Section 2.3.3, while the case of rotational drives is described in Section 2.3.2. In both the approaches, a laminate structure composed of an elastic substrate and a piezoelectric ceramic is driven in resonance to produce (through superposition of sinusoidal perpendicular

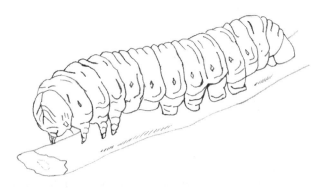

Figure 1.16   Biological model for inchworm piezoelectric actuators.

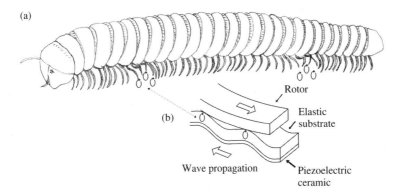

Figure 1.17   Biological model for travelling wave ultrasonic linear and rotational motors.

motions) an effective elliptic displacement at the interface between rotor and stator (see Figure 1.17b). This provides traction through a frictional transmission.

1.5   Concomitant actuation and sensing: smart structures

In previous sections, we discussed how transducing devices can be configured either as sensors or as actuators. Here, we stress the fact that transducers can implement both crucial roles – sensing and actuation – in a motion control system. The former provides a tool for monitoring the status of a system and the latter enables us to impose a condition on a system.

In some special cases, transducers can be used both as sensors and as actuators. Take for instance, DC electromagnetic motors. Rotational electromagnetic motors can function both as actuators (i.e. rotational mechanical energy is produced by the application of a driving voltage), and as generators (i.e. the rotation of the output shaft produces a voltage, proportional to the rotation rate, at the electrical terminals).

Let us consider the case of a permanent magnet DC electromagnetic motor. DC electromagnetic actuators make use of Lorentz's electromagnetic interaction between a permanent magnetic flux, B, and an electrical current, i, flowing in a coil (see Figure 1.3). In an appropriate configuration, if the coil is allowed to rotate, the magnetic interaction produces a torque, T, on the coil, causing rotation. The torque developed by the electromagnetic interaction can be expressed as:

$$T = k_T i \tag{1.13}$$

where k_T is the so-called torque constant of the DC motor.

Similarly, since the DC motor coil is rotating in a magnetic field, induction will take place and a (back) electromagnetic voltage (EMF) will be induced. The expression for the back EMF is:

$$V_{EMB} = k_V \omega \tag{1.14}$$

where ω is the angular rate of the coil and k_V is the voltage constant or back-EMF constant of the motor.

When the DC electromagnetic motor is driven by an input voltage, an angular rate will be developed. It can be demonstrated, in a first approximation, that the angular velocity of the motor obeys the following differential equation when no external load is applied to the shaft:

$$\frac{d\omega}{dt} = -\frac{1}{J}k_F \omega(t) + \frac{1}{J}k_M i(t) \tag{1.15}$$

where J is the motor's rotational inertia and, k_F is the viscous damping constant equivalent to the frictional forces in the motor.

Likewise, when no voltage is applied to the DC motor terminals and the shaft is rotated at a constant velocity, ω, a voltage will be developed between the terminals according to Equation 1.14. This means that the motor works either as an actuator or as a sensor.

A transducer cannot be operated both as a sensor and as an actuator simultaneously unless a model of the transduction is available; in other words, the device can only be used for one of its functions at a time. For practical purposes, this means, in the case of the previous example, that if a rotational velocity is being imposed by means of a DC electromagnetic motor, the same motor cannot be used to sense the rotational velocity that is being imposed.

The previous discussion is true and holds for all transducers; however, some sensing and actuation functions can still be implemented concomitantly. Equations 1.13 and 1.14 can be rewritten according to the following expression:

$$\begin{bmatrix} V \\ i \end{bmatrix} = \begin{bmatrix} 0 & k_V \\ k_T & 0 \end{bmatrix} \begin{bmatrix} T \\ \omega \end{bmatrix} \tag{1.16}$$

It can be shown that $k_V = -1/k_T$; thus, the motor can be classified as a gyrating transducer. As discussed previously, in gyrating transducers, there is a causal relationship between the effort at one port and the flow at the other port. If the torque (effort) is selected as the independent variable, the current (flow) is causally determined. The motor can be used to sense the load (torque) by monitoring the electrical current.

Piezoelectric actuators are a similar case. A piezoelectric actuator establishes a flow of energy from the electrical to the mechanical domain according to the constitutive equations of the piezoelectric effect (see Chapter 2). When no external load is applied to a piezoelectric stack actuator, the displacement (strain) will be a nonlinear, hysteretic function, $S_1(V)$, of the voltage applied at the input port. Wherever an external force is applied to the actuator, it will act as a disturbance to the output displacement. The complete relationship between strain, voltage and load will take the form of Equation 1.17 and is commonly called an operator-based actuator model of the piezoelectric stack transducer (Kuhnen and Janocha (1998)).

$$S(t) = S_1(V) + kf(t) \tag{1.17}$$

where k is the piezoelectric stack stiffness.

Similarly, the charge developed in the piezoelectric stack, $Q(t)$, will be a direct function of the load applied to the transducer, $f(t)$. This time, the voltage-induced charge during operation will act as a disturbance to the operator-based sensor model described by Equation 1.18.

$$Q(t) = df(t) + Q_1(V) \tag{1.18}$$

where d is the piezoelectric coefficient and $Q_1(V)$ is a nonlinear, hysteretic function of the voltage.

Again, even though the piezoelectric stack cannot be used to impose a displacement (strain) and to concomitantly sense it, the sensor model of Equation 1.18 can be used to estimate the load on the actuator, that is, the piezoelectric stack is being used concomitantly to impose a displacement and to sense the load. The estimated load can then be used to compensate for its disturbing effect on the displacement of Equation 1.17 (Kuhnen and Janocha (1998)).

A model of the transduction process can be used to implement both functions (sensing and actuation) at a time. Before discussing this possibility in detail, let us recall here Equation 1.10, which describes the relationship between effort and flow variables in the electric-circuit analogy:

$$V = Z_e I + T_{em} v \quad \text{and} \quad F = T_{me} I + Z_m v$$

The first equation describes the transducer as an actuator, that is, the application of a voltage, V, leads to a current drawn, I, and to an output velocity, v. The Laplace transform of this first equation is:

$$V(j\omega) = Z_e I(j\omega) + T_{em} U(j\omega) \tag{1.19}$$

The overall electrical voltage includes a term dependent on the current drawn, $Z_e I(j\omega)$, and a term related to the output velocity, $T_{em} U(j\omega)$. This equation indicates that the output velocity could be estimated by measuring the overall voltage, V, and subtracting the voltage drop, V_{Z_e}, across the actuator's blocked impedance, $V_{Z_e} = Z_e I(j\omega)$.

The above result provides the basis for estimation of the actuator's motion from a *bridge circuit configuration*, as shown in Figure 1.18. This result is important in that it could lead to: (i) modification of the actuator's behavior (for instance its damping characteristics) through the implementation of feedback control loops based on the estimation of the velocity and (ii) collocated and concomitant sensing and actuation.

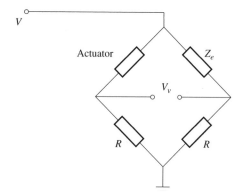

Figure 1.18   Bridge circuit for producing a signal proportional to the actuator's velocity in concomitant sensing and actuation.

If a copy of the actuator's blocked impedance is used in the bridge circuit branch as depicted in Figure 1.18, the voltage across the bridge, V_v, is proportional to the actuator's velocity.

The first approach, that is, modification of the actuator's damping properties, has been studied in the context of voice coil loudspeakers (de Boer (1961)). In this case, the feedback from the unbalanced bridge voltage is utilized to increase damping around the resonance frequencies. The second approach has been implemented in collocated and concomitant position and velocity feedback in piezoelectric actuators (see Dosch *et al.* (1992) and Hagwood and Anderson (1991)).

The main problem in this approach is measurement of the actuator's blocked impedance. It has been found that in most implementations the blocked impedance, rather than being constant and independent of the actuator's motion, is a nonlinear function of the current drawn.

Ideally, if output velocity could be estimated from the voltage across the bridge, the sensing part of the electric-circuit analogy (see Equation 1.20) could then be used to produce an estimate of the mechanical conjugate variable (the force).

$$F(j\omega) = T_{me}I(j\omega) + Z_m U(j\omega) \tag{1.20}$$

In some instances, two-directional transduction of the same conjugated variable can be achieved when a transducing material is used to develop an actuator. This is only possible where the actuation process is accompanied by a concomitant change in any of the material properties of the transducer.

This is true of shape memory alloy actuators (SMAs) (see Chapter 3). In SMAs, thermal input energy is used to promote a phase change in the material. This phase change is accompanied by recovery of the shape induced by deformation. A number of physical properties of the material are altered during the process of shape recovery. In particular, the electrical resistivity of the material is modified by the thermally driven shape change. The electrical resistance can be used to monitor the shape recovery. The actuator can be used to impose a displacement (strain) and simultaneously to sense this displacement.

Concomitant sensing and actuation is a very powerful phenomenon in mechatronic system design (see Section 1.4). As shown above, it allows functions to be shared on a single component. When properly exploited, this can result in very compact and smart solutions.

On the basis of the use of concomitant sensing and actuation is the concept of a *smart actuator*. When a two-directional transducer of this type is embedded in a structure, the combination is usually referred to as a *smart structure*.

The term *active structures* is also commonly found in the literature. The main difference between a smart and an active structure is more a matter of the degree of integration than of the functionality of the embedded smart actuators. In smart structures, a higher degree of integration is assumed: that is, two-directional transducers are highly integrated in the structure so that the whole structure can be considered a functional continuum.

In smart actuators, two functions are combined on a single component. When analyzing smart actuators, particular attention must be paid to the rationale behind the development of monitoring transducers (sensors) and acting transducers (actuators).

As noted earlier, sensors are intrinsically low-power transducers. A minimum energy interaction with the plant must be ensured so that the monitoring process does not interfere with the plant. Miniaturization is therefore a logical and desirable trend.

On the other hand, actuators are intrinsically high-power transducers and as such should impose a state without being disturbed by the plant. In actuators, the trend toward miniaturization is not a logical consequence of their interaction with the plant.

There are two opposite design goals driving the development of sensors and actuators. The former are low-power devices, best approached in a miniaturized fashion, while the latter are intrinsically high-power devices, conceptually the opposite of miniaturized components. These apparently opposing requirements can only be met by high-power density transducing materials. *Power density* in emerging actuators is considered throughout this book as one of the driving forces in the development of new technologies.

1.6 Figures of merit of actuator technologies

Actuators drive plants in motion control systems in obedience to control inputs. They are used to impose the controlled variable on the plant in accordance with the reference trajectory.

Imposing a state on the parameter of the plant raises a number of issues:

1. *Univocal correspondence between input signal and imposed system variable.*
   Ideally, there should be a unique output value corresponding to the system's parameter.

2. *Linearity*. Even though the above univocal correspondence will not generally be linear, linearity is always desirable.

3. *Stability*. The correspondence between input and output should not be influenced by drifts. The intrinsic high-power characteristics of actuators usually leads to thermal drift.

In selecting actuators for a particular application, a number of requirements may arise. These include power or force density, efficiency, size and weight, and cost. The following paragraphs briefly describe the figures of merit of actuators. In some instances, these figures of merit are perfectly quantifiable (e.g. force or power density); in other cases, quantification is not possible, and a thorough analysis of the application and the actuator characteristics for matching will be required.

The different figures of merit are arranged in the following categories: dynamic performance, behavior upon scaling, static performance, impact of environmental parameters, suitability to the application and cost.

1.6.1 Dynamic performance

In general, actuators are used under dynamic operation conditions. Dynamic operation entails continuous changes in the reference trajectory and in the loading conditions imposed by the mechanical interaction with the environment at the output port.

Dynamic operation usually produces changing conditions in the amount of energy flow across the actuator (power requirements), in the relative value of the conjugate variables (velocity and force) at the mechanical port and in the efficiency of transduction between input and output energy.

There are several indicators that can be used to measure the dynamic performance of actuators. Some of these are analyzed in the coming sections.

Power density and specific power density

Power density, P_V, is the ratio of the maximum available mechanical output power, P_{out}, to the volume of the actuator, V:

$$P_V = \frac{P_{out}}{V} \tag{1.21}$$

If the ratio of output mechanical power to the weight, ρV, of the actuator is considered, this defines *specific power density*, P_ρ:

$$P_\rho = \frac{P_{out}}{\rho V} \tag{1.22}$$

Power density and specific power density are measures of the rate of energy delivery at the mechanical port. They are also a measure of how suitable a transducing technology is as a smart actuator that is also for simultaneous use as a sensor.

Work density and specific work density per cycle

Like power density, *Work Density per cycle*, W_V, is defined as the amount of mechanical work that an actuator can deliver during an actuation cycle and is defined by the ratio of output work to volume:

$$W_V = \frac{W_{out}}{V} \qquad (1.23)$$

Likewise, *specific work density per cycle*, W_ρ, is defined as the ratio of maximum available output mechanical work per actuation cycle to the weight of the actuator:

$$W_\rho = \frac{W_{out}}{\rho V} \qquad (1.24)$$

In practice, both power and work densities are difficult to standardize as indicators for dynamic performance. This is basically due to uncertainty as to what should be considered the actuator volume or weight. Taking for example the case of traditional pneumatic linear actuators, if the volume or weight of the pneumatic cylinder is considered, the resulting work density is high as compared, for instance, to electromagnetic drives. However, if the accompanying components (power source, proportional or "on–off" valves, fluid-filtering components) are considered, the situation may be the reverse.

Power and work per cycle density and specific density are related through the actuator's available working frequency, f. Thus,

$$P_V = W_V f \qquad (1.25)$$

$$P_\rho = W_\rho f \qquad (1.26)$$

Time constant and frequency bandwidth

The *Time constant*, τ, of a first-order system is the time taken for the output parameter of the system to reach 63.2% of its final value upon the application of a step input. In actuator systems, the mechanical time constant, τ_m, is usually defined as the time required for the output velocity of the actuator to reach 63.2% of its final value under no external load.

Owing to the inherent power limitations of any actuator system, the frequency response of the actuator will take the form of a low pass filter. This was illustrated in Figure 1.6. The cutoff frequency is defined as the frequency at which a decay of 3 dB in the output velocity of the actuator is observed.

The available bandwidth of the actuator is then defined by the cutoff frequency. Both the time constant and the maximum available frequency of an actuator are related by the following expression:

$$f = \frac{1}{2\pi \tau} \qquad (1.27)$$

Energetic efficiency

The *efficiency*, η, in the transduction process in an actuator is defined as the ratio of the output mechanical energy, W_m, to the input electrical energy, W_e. In most emerging actuators, an extension of the actuator concept will be necessary in order to apply this definition, as the input of the transducer usually is in a nonelectrical domain: for example, magnetostrictive (magnetic domain), SMA (thermal domain), polymer gels (chemical domain).

$$\eta = \frac{W_m}{W_e} \tag{1.28}$$

Ideally, actuators are lossless devices, and, thus, efficiency in an ideal situation should be close to 100%. In practice, various different dissipative phenomena take place in the transducer or accompanying components, producing lower efficiency.

The transduction efficiency of all actuator technologies is a dynamic parameter. In general, the efficiency of the actuator is a function of the actuation conditions. The maximum efficiency is usually taken as the figure of merit.

1.6.2   Actuator behavior upon scaling

Current technological trends towards miniaturization impose strict requirements on actuators. Actuators are intrinsically high-power devices. The higher the power they can deliver, the more optimal their performance is.

Higher-power availability is an indication for instance of higher-frequency bandwidth or higher rejection of load disturbances. Miniaturization does not therefore logically lead to optimization of actuator performance. Rather, miniaturization of actuators must be seen as an application requirement.

The behavior of an actuator upon scaling is a characteristic of each technology and can be assessed by analyzing how the various different performance parameters (efficiency, power and work density, response time, force and stroke) evolve upon scaling.

The analysis of scaling of actuators is a complex task. The reader is referred to Madou (1997) and Peirs (2001) for a detailed study of scaling. Here, we will only give some theoretical background with experimental examples where possible.

For direct transducing operations, finding the available force, stroke and work density upon scaling is straightforward. If we let L be the dominant dimension of the actuator, the following can be said:

- *Force upon scaling.* When analyzing the available force of an actuator, the relevant dimension, L, for most technologies (e.g. piezoelectric actuators, shape memory alloy actuators, magnetostrictive actuators and most electroactive actuators) is the dimension of the cross section. The force, F, is then easily found following the scaling law of Equation 1.29.

$$F \propto L^2 \tag{1.29}$$

Upon scaling the dominant dimension, L, the available force scales as L^2. Dimensions multiplied by 10 lead to available force multiplied by 100. The opposite occurs when scaling down the actuator's dimensions.

- *Stroke upon scaling.* In this case, the stroke, S, of the actuator is usually given as a percentage of its length. Thus, the dominant dimension is the length of the actuator, L. The stroke scales linearly with the scaling of the actuator:

$$S \propto L \qquad (1.30)$$

When the dimensions of the actuator are multiplied or divided by 10, so is the stroke.

- *Work density and specific work density upon scaling.* Work can be readily determined as the product of displacement and force, $W_m = F \cdot S$. In addition, the volume of an actuator obeys a scaling law proportional to the third power of the dominant dimension, $V \propto L^3$. It follows, then, that the work density, defined as the ratio of work to volume, scales according to the following expression:

$$W_V \propto L^0 \qquad (1.31)$$

The above equation indicates that for most actuator technologies, the available work density per cycle remains roughly constant upon scaling.

When considering the effect of scaling on dynamic properties (power density, time constant, frequency), the analysis becomes more complex. This entails identifying what particular factors will become dominant upon scaling, so that they effectively limit the dynamic performance of the actuator. Once the dominant factor is identified, its evolution upon scaling is estimated.

In particular, the time constant of the actuator (which can be used to work out all the other dynamic properties from the static ones) may be limited by a variety of factors for a single actuator technology. In the case of piezoelectric actuators in particular, the time constant (maximum frequency) can be limited by:

1. The resonance frequency of the actuator, which in most cases imposes the driving bandwidth,

2. The heating of the piezoelectric ceramic, which can lead to depolarization if the Curie temperature is reached,

3. The charging time of the capacitor.

In other actuator technologies, the limiting factors for the time response may be very different: heat dissipation (conduction or convection) in thermal actuators; mass transport or diffusion in ionic-type EAPs.

1.6.3   Suitability for the application

The suitability of an actuator technology for a particular application is hard to quantify, but it is usually one of the aspects considered when adopting a particular technology for an application. Suitability may involve a variety of aspects, but it commonly depends on a particular actuation characteristic that is intrinsically matched by the conditions of the application.

Two examples will illustrate this point. Let us first consider the temperature control in a process for mixing two fluids. The resulting fluid must remain between upper and lower temperatures of T_u and T_l respectively. In these conditions, a thermal actuator may be the right choice.

If, for example, an SMA actuator is chosen, it can be made to open and close the hot fluid valve directly in response to the temperature of the mixed fluid in which it is immersed. A similar application to this example is described in more detail in Case Study 3.2, page 135.

For our second example, let us consider conducting polymers in biomedical applications (see Chapter 6). Conducting polymers are soft ionic actuators. In order to actuate, they have to be immersed in an aqueous electrolyte. This requirement is usually a shortcoming rather than an advantage; in most applications, they require packaging solutions to keep the actuator wet during operation.

In the biomedical field, most applications are naturally realized in aqueous electrolytes (blood, urine, etc.). If other actuator technologies are to be applied under these conditions, they must be protected against these corrosive environments. However, it is the ideal environment for CP actuators.

1.6.4   Static performance

The static performance of actuators is typically evaluated on the basis of their available maximum effort (force or torque) and their maximum output velocity or stroke.

Blocking effort

The blocking effort is defined as the maximum effort (force or torque) that the actuator can deliver. This is the effort that will block the actuator so that no further displacement can be achieved against this load.

In the case of rotational actuators, the blocking effort is usually referred to as *stall torque*.

Maximum stroke

The maximum stroke (if any) of an actuator is the maximum available displacement that the actuator can deliver. It is the value of the displacement when no external load is applied on the actuator.

For most emerging actuators, the maximum stroke is given as a percentage of its length. In other actuators (pneumatic and hydraulic), it is limited by the particular configuration of the piston and is given as an absolute value.

Other actuators present no limit on the displacement they can attain. This is true of most rotational actuators (electromagnetic motors, ultrasonic motors, etc.). In these cases, the maximum rotational speed is commonly given.

1.6.5   Impact of environmental parameters

As noted earlier, for optimal use, actuators should be as insensitive as possible to external parameters. These typically involve temperature fluctuations, humidity changes and other external factors.

Temperature and temperature fluctuations have a direct undesired effect on most actuator technologies: there are upper limits on the temperatures that piezo-electric actuators can sustain because of depolarization; the electrical conductivity characteristics of ERF actuators can change as a result of temperature fluctuations.

Humidity has a direct effect on *wet* EAPs, so that there are strict packaging technology requirements unless the application is intrinsically wet.

1.7   A classification of actuator technologies

Actuators, as a particular category of transducers, can be classified according to a variety of criteria. Since the main function of an actuator in a mechatronic system is to establish a flow of energy between an input domain and the output domain, the first category heading is the sign of the power transmission.

Power is assumed to be positive when energy flows from the transducer to the plant and not vice versa. This classification, which categorizes actuators as semiactive or active devices, was introduced in an earlier section.

1.7.1   Semiactive versus active actuators

The power at the output port of the actuator can be expressed as a function of the conjugate variables as:

$$P_{Trans} = F \cdot v \qquad (1.32)$$

for translational output mechanical energy, and:

$$P_{Rot} = T \cdot \omega \qquad (1.33)$$

for rotational mechanical energy.

Semiactive actuators are those whose output mechanical power is not positive: $P_{Trans} \leq 0$ or $P_{Rot} \leq 0$. This means that the energy level in the plant is reduced. Semiactive actuators dissipate the energy of the plant they are coupled to.

Semiactive actuators can actively modulate power dissipation, but the effort they supply (whether a force or a torque) can only oppose the flow in the plant (whether a velocity or an angular rate).

Where semiactive actuators are used in motion control systems, these are known as *Semiactive motion control systems*. They are particular implementations of mechatronic systems in which the objective is to maintain the energy level of the plant within a bounded region. A typical example of a semiactive motion control system is the use of ER or MR fluid actuators for vibration isolation of delicate or fragile equipment from noise sources.

Semiactive control of vibrations, as the paradigmatic application of these systems, is analyzed in detail in Section 6.3. In addition, several instances of application to semiactive vibration isolation are analyzed in Case Studies 6.1 to 6.3.

Active actuators can either increase or decrease the energy level of the plant to which they are coupled. The power flow can be either positive or negative: $P_{Trans} \gtrless 0$ or $P_{Rot} \gtrless 0$.

Of the various different emerging actuators discussed in the book, only ERF and MRF actuators are classified as semiactive.

1.7.2 Translational versus rotational actuators

The conjugate variables used to define the power output in a translational actuator are the linear velocity, v and the force, f. The conjugate variables for rotational actuators are the angular rate, ω, and the torque, T.

Translational actuators convert electrical energy into translational mechanical energy, while *rotational actuators* convert electrical energy to rotational mechanical energy.

Rotational actuators are always obtained from geometrical transducers, for example, an electromagnetic DC motor. On the other hand, transducing materials generally produce translational actuators, for example, magnetostrictive actuators, unless some geometrical concept is added.

Depending on the type of motion resulting from the transduction process, *bending actuators* may also be considered (for instance, piezoelectric multimorph actuators). However, bending actuators are most often used in the context of driving linear plants.

In general, actuators of both types can be developed from geometrical concepts for all transducing materials. This applies particularly to piezoelectric actuators. Piezoelectric actuators lead, through different geometrical concepts, to rotational, translational and bending actuators.

1.7.3 Input energy domain

The classification according to the transduction principle (see Section 1.2) gives an idea of what classification according to input energy domain is like. Here, the

Table 1.2 Summary of actuator classification.

Type	Power flow		Output energy		Input energy domain					Force flow	
	Active	Semiactive	Linear	Rotational	Electric	Thermal	Magnetic	Chemical	Fluid	Soft	Hard
Piezoelectric actuators											
TWUM	X			X	X						X
TWLUM	X		X		X						X
Stacks	X		X		X						X
Inchworm	X		X		X						X
Multimorph	X		Bending		X						X
Shape memory actuators											
Mass load	X		X	X	X	X				X	
Spring load	X		X	X	X	X				X	
Antagonistic	X		X	X	X	X					X
Wet EAP actuators											
Polymer gels	X		X		X	X		X		X	
IPMC actuators	X			Bending	X					X	
CP actuators	X			Bending	X					X	
Carbon nanotubes	X		X		X					X	
Dry EAP actuators											
Electrostrictive	X		X		X					X	
Dielectric elastomers	X		X		X					X	
Field responsive fluid actuators											
MRF actuators		X	X	X	X		X		X		X
ERF actuators		X	X	X					X		X
Magnetostrictive actuators											
Magnetostriction	X		X				X			X	
MSM actuators	X		X				X			X	

output energy domain is restricted to the mechanical domain, without any further separation of energy types (rotational and translational).

There are five main input domains:

1. *Input electrical energy.* Most actuators belong to this category, particularly all piezoelectric actuators, electrostatic actuators, dry EAPs and ERF actuators.

2. *Thermal electrical energy.* This category includes SMA actuators, some wet EAP actuators (in particular, some Polymer Gels) and thermal bimetallic actuators.

3. *Magnetic electrical energy.* This category includes magnetostrictive actuators, MRF actuators and magnetic shape memory (MSM) actuators.

4. *Chemical input energy.* This category includes some wet (ionic) EAPs.

5. *Fluid input energy.* The pressure of a fluid in a chamber is used to provide the actuator force. This category includes *pneumatic* and *hydraulic* actuators. This domain, together with the thermal domain, is considered a particularization of the mechanical energy domain.

1.7.4 Soft versus hard actuators

Soft actuators, also called *pulling actuators*, are based on transducing materials configured in thin sheets or wires so that they can only withstand traction forces. The operative principle restricts the available forces to pulling:

$$f_{Soft} \geq 0 \qquad (1.34)$$

Hard actuators, also known as *push–pull actuators*, have the ability to sustain both traction and compression forces:

$$f_{Hard} \gtrless 0 \qquad (1.35)$$

Hard actuators are inherently two-directional actuators. Soft actuators are inherently unidirectional actuators but can be configured in antagonistic pairs to provide two-way actuation. This is commonly true of SMA actuators.

There are other possible classification criteria for actuators and transducers. For instance, actuators can also be classified according to whether the output motion is continuous or discontinuous. A classical example of discontinuous operation is electromagnetic or piezoelectric steppers. Table 1.2 summarizes the classification convention followed in this book.

1.8   Emerging versus traditional actuator technologies

Traditional actuators have been employed extensively during the last century in all application domains. The category of traditional actuators includes three main technologies, namely, *Electromagnetic motors, pneumatic actuators* and *hydraulic actuators.*

Electromagnetic motors exploit Lorentz's interaction between an electrical charge and a magnetic field in which it moves. In the case of a rotational motor, the transducer equation for this technology is

$$\begin{bmatrix} V \\ i \end{bmatrix} = \begin{bmatrix} 0 & g \\ -1/g & 0 \end{bmatrix} \begin{bmatrix} T \\ \omega \end{bmatrix} \tag{1.36}$$

These may be considered as gyrating transducers. In general, the applied voltage, V, will determine the rotational or translational velocity, ω or v, respectively. The current drawn is an indication of the torque or force applied at the mechanical port.

There are many different types of electromagnetic motors, but an exhaustive discussion of this technology lies outside the scope of this book. The reader is referred to any of the countless reference books on motion control hardware.

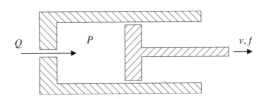

Figure 1.19   Transduction between fluid and mechanical domains, typical of pneumatic and hydraulic actuators.

Table 1.3   Comparison of traditional and emerging actuators.

Traditional actuators	Emerging actuators
Based on geometrical transducers	Based on transducing materials (possibly in combination with geometrical concept)
Off-the-shelf availability	Designed for the application
Good performance at normal scale	Good for meeting miniaturization demands
Lumped approach: discrete components in MCS	Integrated and embedded approach: open to smart structure concepts
Used in combination with external sensors	Pursuit of the smart actuator concept
Conventional mechanical transmissions for (output) impedance matching	New transmission designs based on hinges and friction
Incompatible with biomedical applications	Technologies (in some cases) ideally suited to biomedical applications

Pneumatic actuators exploit the power of a fluid (a gas, usually air) flowing into a chamber to develop a force (see Figure 1.19). The input energy is in the fluid domain. The power is defined by the volume flow rate, Q, and the pressure, P, as conjugate variables.

Hydraulic actuators are equivalent to pneumatic actuators in that the input energy is also in the fluid domain. The fluid is, however, an incompressible liquid (usually oil). For a linear actuator, the equation of the transducer is ideally

$$\begin{bmatrix} P \\ Q \end{bmatrix} = \begin{bmatrix} k & 0 \\ 0 & -1/k \end{bmatrix} \begin{bmatrix} f \\ v \end{bmatrix} \tag{1.37}$$

where k is the effective area of the pneumatic cylinder.

Emerging actuators are those driving technologies developed from novel (or when old, newly developed) transducer materials. They are analyzed in detail throughout this book. Table 1.3 compares traditional and emerging actuators.

1.9   Scope of the book: emerging actuators

This book is devoted to the analysis of emerging actuators as constituents of motion control mechatronic systems, and as mechatronic systems on their own. The electromechanical design, particular control concepts and exploitation of the smart actuator approach is therefore analyzed for each technology.

The analysis is divided into five chapters dealing with five transduction technologies and the actuator concepts based on these technologies. The transducing materials are the following:

1. *Piezoelectric ceramics*, Chapter 2.

2. *Shape memory alloys*, Chapter 3.

3. *Electroactive polymers*, Chapter 4.

4. *Magnetostrictive materials*, Chapter 5.

5. *Electro- and magnetorheological fluids*, Chapter 6.

The following aspects are addressed for each transducing material:

- detailed description of transduction principles and characteristics;

- analysis of constitutive equations for the transducer and the actuator concepts;

- mechatronic aspects of actuator design;

- control aspects of particular relevance for each technology;

- figures of merit and scaling properties; and

- relevant illustrative applications.

The last chapter of this book summarizes the most salient issues of each emerging actuator technology and presents the comparative position of the various different actuator technologies. Here, traditional actuators are discussed for reference purposes. Particular emphasis is placed on trends in applications and on open research issues.

1.10   Other actuator technologies

1.10.1   Electrostatic actuators

Electrostatic actuators are relatively novel devices whose operating principle is based on electrostatic attractive and repulsive forces between electrical charges. As such, they exhibit the same operating principle as Dielectric Elastomer actuators (see Chapter 4).

Basics of electrostatic interaction and actuators

The basic configuration of an electrostatic actuator is that of a capacitor of variable capacitance, C. The capacitance of a flat, parallel plate capacitor is:

$$C = \varepsilon \frac{A}{d} \tag{1.38}$$

The constitutive equations for the electrostatic actuator can be developed from the expression of the electrical power, P, stored in the capacitor. The power can be obtained from the time rate change of the stored electrical energy, $P = \mathrm{d}W/\mathrm{d}t$. The electrical energy for a capacitor with an applied electric field E is

$$W = \frac{q^2}{2C} = \frac{1}{2}\varepsilon E^2 V = \frac{1}{2}\varepsilon E^2 ayx \tag{1.39}$$

where q is the electrical charge at the capacitor plates, V is the volume of the dielectric between the plates and the other dimensions are as in Figure 1.20.

If a variable gap capacitor is considered, the resulting expression for the power will be

$$P = \frac{\mathrm{d}W}{\mathrm{d}t} = \frac{\partial W}{\partial x}\frac{\mathrm{d}x}{\mathrm{d}t} + \frac{\partial W}{\partial q}\frac{\mathrm{d}q}{\mathrm{d}t} \tag{1.40}$$

In Equation 1.40, the term $\mathrm{d}x/\mathrm{d}t$ is the velocity of the moving plate and $\mathrm{d}q/\mathrm{d}t$ is the electrical current used to charge the capacitor. Since the result of the two terms at the right-hand side of the equation must be a power, it follows that $\frac{\partial W}{\partial x}$ must be the electrostatic force on the moving plate, $F_\perp$, and $\frac{\partial W}{\partial q}$ must be the voltage across the capacitor, U.

If we use Equation 1.39 together with the previous results, it follows that

$$F_\perp = \frac{\partial W}{\partial x} = \frac{1}{2}\varepsilon E^2 ay \tag{1.41}$$

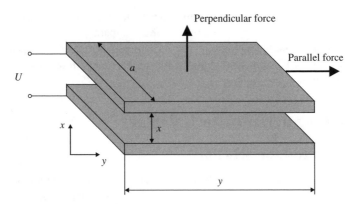

Figure 1.20   Schematic representation of a variable capacitance electrostatic actuator.

and

$$U = \frac{\partial W}{\partial q} = \frac{q}{C} \tag{1.42}$$

Here, let us consider a case in which the variable capacitance configuration is obtained by a modification of the capacitor's effective area while maintaining the gap. If this is achieved by sliding one of the plates sidewise in the direction of y (see Figure 1.20), the expression for the force is

$$F_{\shortparallel} = \frac{\partial W}{\partial y} = \frac{1}{2}\varepsilon E^2 ax \tag{1.43}$$

Equations 1.41 and 1.43 can be rewritten in terms of the applied voltage:

$$F_{\shortparallel} = \frac{1}{2}\varepsilon U^2 \frac{a}{x} \quad \text{and} \quad F_{\perp} = \frac{1}{2}\varepsilon U^2 \frac{ay}{x^2} \tag{1.44}$$

A simple inspection of Equation 1.44, will readily show the following.

1. *Electrostatic interaction is short range.* The magnitude of the electrostatic forces decrease rapidly when the actuator is scaled up.

2. *Perpendicular forces are much higher than parallel forces.* In most, $F_{\perp} \approx 10^3 F_{\shortparallel}$.

3. *Electrostatic interaction is appropriate for microapplications.*

4. *Complementarity in force and stroke.* Actuators based on perpendicular and parallel forces are complementary in force and stroke. In perpendicular actuators, the stroke is limited at most to the size of the gap.

Actuator configurations

Electrostatic actuators exploit either perpendicular, $F_\perp$, or parallel forces, $F_{\parallel}$. In parallel configurations, the most common developments are the ones known as *comb drives*.

In comb drives, two comblike structures are used. The pins in each comb structure are used as electrodes in a layered capacitor configuration. The parallel-type electrostatic forces tend to pull both combs together. The combination of multiple layers and the symmetry of the structure is used to balance out the perpendicular interaction so that $F_\perp$ on each pin would ideally cancel out.

Comb drives can be used to develop both linear and rotational actuators. Figure 1.21 illustrates the two concepts. In both cases, the parallel force generated by the electrostatic interaction results in interdigital rotational and translational movements.

The other typical configuration for electrostatic actuators is one that exploits perpendicular forces. This configuration is illustrated by the tilt mechanism of microscope mirrors as depicted in Figure 1.22. The mirrors can be rotated around a torsional hinge (light white in Figure 1.22b) as a consequence of the electrostatic force of perpendicular force actuators.

1.10.2 Thermal actuators

Thermal actuators are based on the thermal expansion of materials, either in solids or gases. SMA actuators (see Chapter 3), and some polymer gel actuators (see Chapter 4), can be considered thermal actuators as well as actuators based on thermal expansion.

However, the term thermal actuators usually refers to multimorph-type thermally activated actuators. These actuators exploit the difference in thermal expansion coefficients of two dissimilar metals bonded together to produce bending deformations of the composite structure.

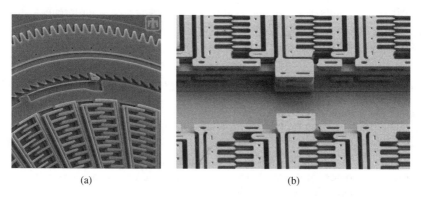

(a)   (b)

Figure 1.21   Electrostatic comb drive actuators: (a) Rotational actuator in combination with a ratchet and (b) detailed view of a linear actuator structure.

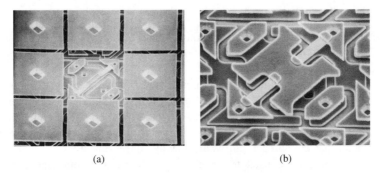

(a)   (b)

Figure 1.22   Matrix of controlled tilt mirrors for optical deflection. Each pixel is 16 μm in width and is mounted on electrostatic actuators.

Thermal actuators are limited by the long response time of the heat transport process required for actuation. The force available in a thermal actuator is proportional to L^2, where L is the dominant dimension of the actuator. Similarly, the available stroke is usually expressed as a percentage of the actuator length and thus scales as L.

From the force and stroke, it follows that the available work per cycle and the available work density per cycle scale in proportion to L^3 and L^0 respectively.

In thermal systems, the time taken to transport the heat and actuate the system is proportional to the mass of the actuator, L^3, and inversely proportional to the heat rate. According to Peirs (2001), the heat rate scales proportionally to L. Therefore, the time constant of thermal systems scales in proportion to L^2.

This means that reducing the scale of the actuator by a factor of 10 will produce systems that are 100 times faster. As calculated above, the power density is proportional to L^{-2} and thus also increases by a factor of 100 when the actuator is scaled down by a factor of 10.

All these results suggest that thermal actuators are an appropriate technology for integration in microsystems. Figure 1.23 shows a thermal actuator in a microsystems application.

1.10.3   Magnetic shape memory actuators

A martensitic transformation is a first-order, diffusionless transformation exhibited by some alloys (Fe, Cu and Ti alloys). The martensitic transformation is affected by a variety of external fields (temperature, uniaxial stress, hydrostatic pressure and magnetic fields) and is the basis of various different emerging actuator mechanisms. In the context of thermally triggered martensitic transformations in Ti-based alloys in particular, we would highlight the case of shape memory alloy actuators (see Chapter 3 for a detailed discussion of SMA actuators and martensitic transformations).

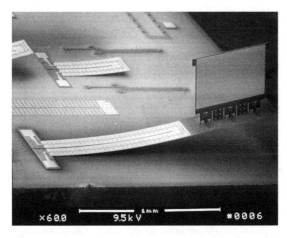

Figure 1.23   SEM photograph of a vertical thermal bimetallic actuator with integrated micromirror. The application of a current to the actuator arm produces vertical motion of the mirror, which can either reflect an optical beam or allow it to be transmitted (Photograph courtesy of Joseph N. Mitchell, Southwest Research Institute).

In this section, we will briefly address the case of actuators based on martensitic transformations triggered by magnetic fields, which are known as *ferromagnetic shape memory alloy actuators, FSMAs* or *magnetic shape memory actuators, MSMAs*. As in the case of thermally triggered shape memory actuators, the transformation is possible because of the difference in magnetic moment between the parent and the martensite variants.

The mechanism of actuation with MSMAs is similar to the mechanism with thermal shape memory actuators. The ferromagnetic shape memory alloy consists of internal domains and twin variants, which have different crystallographic and magnetic orientations. In the martensite phase, when no external magnetic field is applied, the twin variants are preferentially aligned (by means of a bias load). The shortest, magnetically active axis in the crystallographic lattice is generally oriented in the direction of actuation (see Figure 1.24, left).

When a magnetic field is applied perpendicular to the magnetically active axis, the other twin variants appear and grow. As a direct consequence, the magnetically active axis (which is the shortest in the lattice) of these new variants is aligned perpendicular to the direction of actuation and the length of the specimen increases (see Figure 1.24, right). The maximum attainable stroke is defined by the ratio of the length of the long to the short axis in the lattice, a/c.

For the magnetic shape memory effect to materialize, the magnetic anisotropy energy of these alloys must be larger than the elastic or frictional energy associated with the conversion of variants. As in the case of thermally triggered shape memory actuators, MSMAs are one-way actuators and must be configured against bias

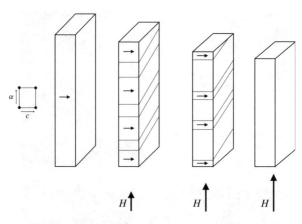

Figure 1.24   Schematic representation of the magnetic shape memory mechanism.

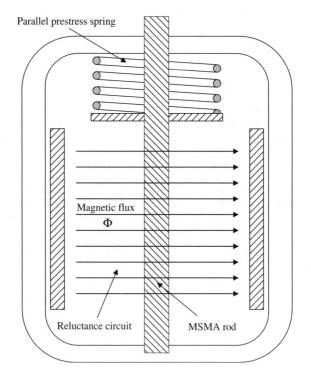

Figure 1.25   Components in MSMA: bias loading springs, MSMA rod and reluctance circuit.

loading in order to complete the actuation cycle. Again, magnetic SMAs can be made to change their shape in response to axial, bending or torsional loading.

The basic structure of a magnetic shape memory actuator is depicted in Figure 1.25. The actuator includes components to prestress the magnetic SMA specimen (usually based on linear springs as shown in the figure) and components to set up the magnetic field perpendicular to the direction of actuation. The reader is referred to Chapter 6 for more details regarding the design of reluctance circuits to set up the magnetic field.

Magnetic shape memory actuators are a special class of giant magnetostrictive actuators, which can therefore be classified into two broad categories: Joule magnetostrictive actuators and twin-induced magnetostrictive actuators. On the one hand, magnetostriction based on the Joule effect has been known since the nineteenth century and is dealt with in detail in Chapter 5. On the other hand, the property known as twin-induced magnetostriction was discovered much more recently (in the 1960s, see Kakeshita and Ullakko (2002)) and is only now gaining some momentum.

Joule-based magnetostriction is characterized by higher forces (only a few N have been achieved in MSMAs so far) and frequency of actuation, higher Curie temperatures (about 380 C as compared to only 100 C in twin-induced magnetostrictive materials) and smaller driving magnetic fields (usually half that required in MSMAs), fatigue and hysteresis.

Twin-induced magnetostriction is characterized by a higher stroke (up to 50,000 ppm, as compared to 1700 ppm in Terfenol–D, see Chapter 5) and higher energy output per cycle (three times that of Terfenol–D). The first prototypes of MSMAs were not developed until recently; they are currently at a promising experimental stage. Some of these prototype implementations will be introduced in more detail in a case study in Chapter 5.

2

Piezoelectric actuators

Actuators based on piezoelectric materials probably represent the most mature and best established of the different emerging technologies. The physical phenomenon of piezoelectricity provides the foundation for the transduction process upon which all piezoelectric actuators are based.

The category of piezoelectric actuators encompasses a number of different configurations. It is common to find both stepping and continuous piezoelectric actuators, rotational and translational drives, micropositioning piezoelectric stages with sub-nanometer resolution, large stroke positioning stages and also fast drives.

This chapter provides an analysis of piezoelectric actuators. After a short historical note, there is an introduction to the basics of piezoelectricity, piezoelectric materials and their constitutive equations.

These driving technologies are classified for analysis into resonant and nonresonant piezoelectric actuators. Various different configurations within these two categories are discussed and the main driving characteristics are highlighted.

In line with the mechatronic focus of this book, we highlight the possibility of using piezoelectric actuators concomitantly as sensors, in the context of developing control strategies for achieving optimal operation points in resonant drives and reduced hysteresis in non resonant drives.

A whole section is devoted to a comparative analysis of the various different piezoelectric actuator configurations. In particular, there is a detailed note on the scaling properties of this technology.

The last part of this chapter addresses the application of piezoelectric actuators, through an analysis of five case studies. Two of these concern resonant piezoelectric actuators, presenting the development of OEM drives and the integration of ultrasonic motors in the optical automatic focus of reflex cameras.

The three last application examples concern nonresonant drives. Here, three very different application domains are analyzed, which lead to three different actuator configurations: piezoelectric benders for needle selection modules in knitting

Emerging Actuator Technologies: A Micromechatronic Approach J. L. Pons
© 2005 John Wiley & Sons, Ltd

equipment; piezoelectric stepping motors for ultrastiff positioning in machine tools; and lastly, modified piezoelectric stack actuators for precise scanning in atomic force microscopy in spacecraft applications.

2.1 Piezoelectricity and piezoelectric materials

Piezoelectricity was first observed in some natural materials (quartz, tourmaline and Rochelle salt), which exhibited electrical polarization as the result of an applied mechanical load. Historically, piezoelectricity was observed by Pierre and Jacques Curie in 1880. The mathematical formulation of the governing constitutive equations for the piezoelectric effect were then developed in the decades following this discovery. The first engineering application of piezoelectricity was devised by Langevin, who laid the foundations of ultrasonic submarine detection in 1916.

The *piezoelectric effect* can be described as modification of the polarization of a dielectric arising from the mechanical energy of the stress. Materials exhibiting such an effect are said to be piezoelectric materials. The piezoelectric effect is reversible in the sense that when an electric field is applied, a mechanical strain will arise. This is the *converse piezoelectric effect*.

In what are called *dielectric* materials, the application of an external electric field leads to electric dipoles. This phenomenon is known as electric polarization and is usually characterized by the vector magnitude $\vec{P}$. In these materials, the electrical charge per unit area is known as electric displacement and is denoted by the vector magnitude $\vec{D}$. The electrical state of a dielectric is then determined according to the following expression:

$$\vec{D} = \epsilon_0 \vec{E} + \vec{P}$$

$$= \epsilon \epsilon_0 \vec{E} \tag{2.1}$$

where ϵ_0 is the vacuum electric permittivity and ϵ is the relative permittivity of the dielectric material.

When the crystal structure of the dielectric material is such that in the absence of external electric fields the centers of positive and negative electrical charges do not coincide, the material is said to exhibit *spontaneous polarization*. If this polarization can be modified when an external electric field is applied, the material is said to be *ferroelectric*.

Ferroelectric materials are not generally piezoelectric. This is mainly because they are isotropic materials; that is, there is no dominant direction for polarization. In order to serve as piezoelectric materials, ferroelectric materials must be subject to an external electric field, the *poling field*.

Therefore, it is the poling process that is responsible for the appearance of piezoelectric properties in ferroelectric materials. This process can take place either at room temperature, once the ferroelectric material is obtained, or in what is called the *paraelectric phase*. The latter approach involves sustained application of an external electric field while the material is cooled to below the Curie temperature, so that the resulting ferroelectric domains have a dominant orientation for the polarization. The Curie temperature becomes the upper limit for the operational range of piezoelectric ceramics; polarization is lost for higher temperatures (see Figure 2.1).

Most current piezoelectric materials are ceramics with a crystalline structure of the *perovskite* type, **ABO₃**. In particular, the well-known PZT and PLZT ceramics are solid solutions obtained from Ti, Zr and Pb oxides. A classic example of *perovskite*-type materials is barium titanate, **BₐTᵢO₃** (see Figure 2.2). In this particular case, the $\mathbf{T_i^{4+}}$ is slightly shifted toward one of the six faces of the cubic structure. When the ceramic is poled, a dominant direction for the shift of the $\mathbf{T_i^{4+}}$ will develop, and there will be a piezoelectric effect.

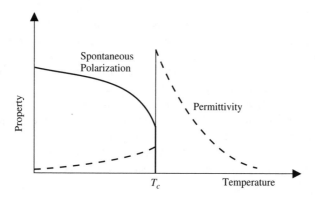

Figure 2.1 Relationship of relative permittivity and spontaneous polarization with temperature: spontaneous polarization in ferroelectric materials is lost at more than the Curie temperature, T_c.

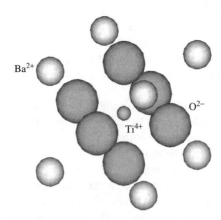

Figure 2.2 Crystal structure of **BaTiO₃**, a classic piezoelectric material.

2.2 Constitutive equations of piezoelectric materials

The functional relation between applied electric field and corresponding material strain in a piezoelectric PZT ceramic can be depicted graphically as shown in Figure 2.3.

When low-amplitude electric field cycles are applied to the piezoelectric materials, a quasilinear mechanical strain is generated. This is the converse piezoelectric effect introduced above, which is characterized by low-level hysteretic behavior. When the applied electric field is increased, the hysteresis level is also increased until there is a *coercitive electric field*. At this level, a degenerated butterfly-like relationship develops between the applied electric field and the resulting mechanical strain.

Piezoelectric ceramics are classified into soft and hard ceramics, according to the relative value of the coercitive electric field:

- *Hard Piezoelectric Ceramics* are characterized by relatively high coercitive electric fields ($\geq$20 kV/cm), a relatively wide linear piezoelectric region, moderate piezoelectric coefficients and high quality factors, Q.

- *Soft Piezoelectric Ceramics* exhibit low coercitive electric fields ($\approx$14–16 kV/cm), relatively higher electrically driven mechanical strain and lower hysteresis levels.

When dealing with actuator applications, particularly in the case of resonant-type piezoelectric actuators, hard piezoelectric ceramics are preferred. This is basically due to the higher quality factor. High quality factors are an indication of high efficiency in the transduction between electrical and mechanical energy, the basis for actuator design.

The quasilinear, low hysteresis piezoelectric effect can be mathematically described by what are known as *constitutive equations*. The constitutive equations

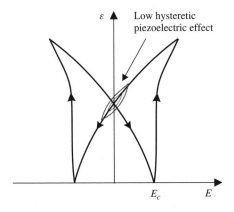

Figure 2.3 Strain–electric field relation for a typical piezoelectric ceramic.

for the piezoelectric effect are described in terms of the relationship between electrical and mechanical variables.

The formulation of the piezoelectric constitutive equations can be derived from a number of thermodynamic energy density functions involving both elastic and electrical energy (Rosen *et al.* (1992)). The particular set of constitutive equations depends on the choice of dependent and independent variables. As we saw earlier in Equation 2.1, a complete definition of the electrical state in the dielectric requires the specification of two independent variables of the set $\{\vec{E}, \vec{D}$ and $\vec{P}\}$. In the following equations, $\{\vec{E}\}$ has been taken as the electrical independent variable and $\{\vec{D}\}$ as the electrical dependent variable.

For the case of elastic energy, the strain, S, and the stress, T, are commonly used. If strain and electric field are selected as independent variables and the electric displacement, $\vec{D}$, is selected as the dependent electrical variable, the constitutive equation of the piezoelectric actuator can be expressed in tensor notation as follows:

$$S_{ij} = c_{ijkl}^E T_{kl} + d_{mij} E_m$$

$$D_k = d_{kij} T_{ij} + \epsilon_{km}^T E_m \tag{2.2}$$

The above tensor notation can be compacted into a matrix notation following the contraction criteria described in Table 2.1. When this contraction criterion is followed, directions 1, 2 and 3 generally represent normal strains, while directions 4, 5 and 6 represent shear strain. In addition, direction 3 is usually associated with the poling direction. See Figure 2.4 for a schematic representation of the axes convention.

Table 2.1   Equivalence between tensor and compact matrix notation.

Tensor notation (ij, kl)	11	22	33	23, 32	31, 13	12, 21
Matrix notation (p, q)	1	2	3	4	5	6

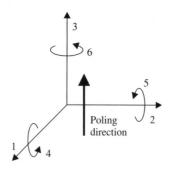

Figure 2.4   Convention for the notation of axes in a piezoelectric ceramic: direction "3" is defined as the poling direction.

In general, according to the rank of all the tensors in the constitutive equations, the electromechanical relationship in a piezoelectric will be completely specified by 21 independent elastic coefficients, 18 independent piezoelectric constants and 6 independent dielectric constants. However, poled ferroelectric ceramics exhibit crystal symmetry, so that there is a reduced set of independent elastic, piezoelectric and dielectric coefficients. In the case of poled ferroelectric ceramics in particular, only 5 independent elastic constants, 3 piezoelectric coefficients and 2 dielectric constants fully describe the material's electromechanical behavior.

In compact notation, the full set of constitutive equations can be described using the following expression:

$$\begin{Bmatrix} S \\ D \end{Bmatrix} = \begin{bmatrix} [s] & [d]^T \\ [d] & [\varepsilon] \end{bmatrix} \begin{Bmatrix} T \\ E \end{Bmatrix} \tag{2.3}$$

where,

$$[s] = \begin{bmatrix} s_{11} & s_{12} & s_{13} & 0 & 0 & 0 \\ s_{12} & s_{22} & s_{23} & 0 & 0 & 0 \\ s_{13} & s_{23} & s_{33} & 0 & 0 & 0 \\ 0 & 0 & 0 & s_{44} & 0 & 0 \\ 0 & 0 & 0 & 0 & s_{55} & 0 \\ 0 & 0 & 0 & 0 & 0 & s_{66} \end{bmatrix} \tag{2.4}$$

$$[d]^T = \begin{bmatrix} 0 & 0 & d_{31} \\ 0 & 0 & d_{31} \\ 0 & 0 & 0 \\ 0 & d_{15} & 0 \\ d_{15} & 0 & 0 \\ 0 & 0 & 0 \end{bmatrix} \tag{2.5}$$

and

$$[\varepsilon] = \begin{bmatrix} \varepsilon_{11} & 0 & 0 \\ 0 & \varepsilon_{22} & 0 \\ 0 & 0 & \varepsilon_{33} \end{bmatrix} \tag{2.6}$$

2.3   Resonant piezoelectric actuators

2.3.1   Basics of resonant operation of piezoelectric loads

Electromechanical characterization

The resonant operation of a piezoelectric motor can be represented by the electrical parameters of the equivalent electrical circuit shown in Figure 2.5. C_0 is the *clamped capacitance* of the piezoelectric vibrator, that is, the capacitance of the piezoelectric ceramic under the boundary conditions of fixed electrodes. Parallel to the clamped capacitance is the motional impedance, which consists of the

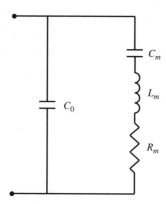

Figure 2.5   Equivalent electrical circuit for the resonant piezoelectric actuator.

motional capacitance C_m, the motional inductance L_m and the resistance R_m. R_m accounts for the power losses within the piezoelectric vibrator, friction losses at the rotor–stator interface (if any) and the mechanical output power delivered to the shaft.

In resonant-type piezoelectric motors, in order to get the maximum mechanical output power, the excitation frequency is tuned to the resonance or antiresonance frequency of the device. A piezoelectric ceramic, as a mechanical continuum, is electrically characterized by an infinity of resonance modes. See Figure 2.6a, b and c for the electromechanical characterization of resonance modes of a piezoelectric motor in a frequency range of 35 to 70 kHz.

A resonance mode is characterized electromechanically by a local minimum of both the electrical and the mechanical impedance. While the electrical impedance is defined as the vector ratio (magnitude and phase) of applied voltage to current drawn, the mechanical impedance is defined as the ratio of vibration velocity to applied exciting force.

As a vector magnitude, the electrical impedance of a piezoelectric actuator depends on its magnitude and phase along the frequency axis. Figure 2.6 shows the electromechanical characterization of a resonant-type piezoelectric motor. The figure shows the particular electromechanical characterization for the operational frequency range.

Figure 2.6a shows the magnitude of the electrical impedance as a function of frequency. Every mechanical vibration mode (see Figure 2.6c) is electrically characterized by a resonance and an antiresonance mode. The resonance mode exhibits a local minimum of the magnitude of the impedance, while the antiresonance mode represents a local maximum of the magnitude (see Figure 2.6a).

Resonance and antiresonance modes are also characterized by abrupt changes in the phase of the electrical impedance. The electrical equivalent of a piezoelectric ceramic is approximately a capacitive load in a frequency range outside the

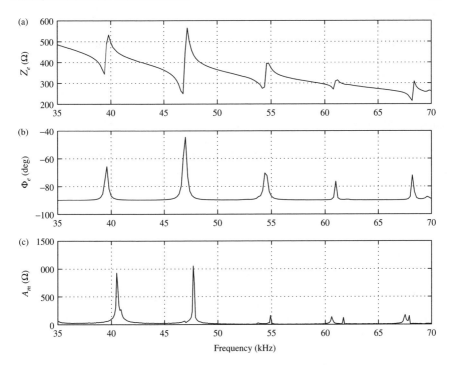

Figure 2.6   Electrical and mechanical impedance curves for the stator of a resonant piezoelectric motor.

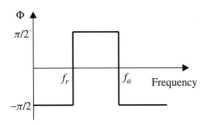

Figure 2.7   Theoretical phase change over the resonance–antiresonance frequency range for a piezoelectric ceramic.

resonance range. In the vicinity of a resonance–antiresonance mode, the piezoelectric load shows electrical behavior proximate to a square-well phase angle function (see Figure 2.7). The phase takes values of $\pm\pi/2$, with discontinuity at the resonance and antiresonance frequencies. In other words, the electrical equivalent in the frequency range between resonance and antiresonance is inductive, and hence the piezoelectric load represents a resistive load at resonance and antiresonance.

The quality factor, Q

One important electromechanical parameter of mechatronic systems is the *quality factor*, Q. The quality factor is a measure of the actuator's efficiency in transduction from electrical to mechanical energy. The quality factor is a magnitude closely related to the shape of the resonance and antiresonance peaks. It is also related to the intrinsic structural damping of the piezoelectric ceramic. In particular, the sharper and the narrower the resonance (the antiresonance) peak, the higher will be the quality factor and the lower the system damping.

The quality factor can be practically derived from the motional impedance of the resonant piezoelectric ceramic. In particular, a good approximation of the quality factor is given by:

$$Q = \frac{\sqrt{L_m/C_m}}{R_m} \tag{2.7}$$

Piezoelectric transducers differ in their operation point according to whether they are conceived as sensors or as actuators. Resonant piezoelectric actuators are usually driven in a frequency range close to the antiresonance frequency of the device. As proposed by Uchino, Uchino and Hirose (2001), antiresonance modes (B-type modes) are better for actuators since their quality factor is higher than A-type modes (resonance modes) over the entire range of vibration frequency (see Figure 2.8).

As a consequence, the driving situation for resonance piezoelectric actuators is as shown in Figure 2.9. The lower limit of the range of frequencies used to drive

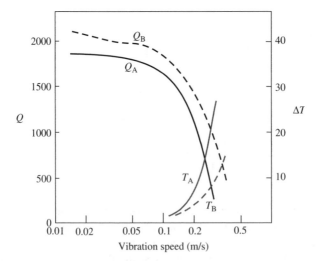

Figure 2.8 Quality factor and heat losses in piezoelectric ceramic as a function of vibration speed. Note the superior performance of B-type modes (antiresonance modes) due to the higher comparative quality factor. Courtesy of K. Uchino.

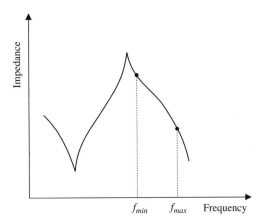

Figure 2.9   Range of frequencies for resonant drives in the region of smooth impedance decay after antiresonance.

the actuator is usually set at a frequency close to antiresonance. The smoothly decreasing impedance curve after antiresonance is therefore used to define the range of frequencies allowed for motor speed control.

Frictional transmission in resonant drives

Resonant-type piezoelectric motors can be classified according to a number of criteria. Several of these classifications are briefly introduced in Section 2.3.2. In this section, we shall focus on the way the microscopic resonant vibration of the piezoelectric ceramic is transmitted and transformed into a macroscopic linear or rotary motion.

In general, the transmission of microscopic vibrational motion into macroscopic motion is based on a friction mechanism between the motor stator and rotor. This is usually achieved in two steps: first, the vibrational motion is transformed to a microscopic elliptic motion of the stator; then, this elliptic motion is transmitted to the rotor through friction or impact mechanisms.

The first step is usually based on a combination of two linear motions on orthogonal axes. The combination of harmonic motion in two mutually orthogonal directions leads to the well-known *Lissajous loci*. Joules Lissajous (1822–1880) was the first to observe and describe these curves during his experiments on optics of vibrational movements. They are obtained by superposition of orthogonal movements of the form:

$$y = A_y \sin(\omega_y t + \varphi_y) \tag{2.8}$$

$$x = A_x \sin(\omega_x t + \varphi_x) \tag{2.9}$$

The exact shape of the Lissajous loci depends on the particular frequencies of the orthogonal movements, ω_y and ω_x. The condition for a closed Lissajous locus

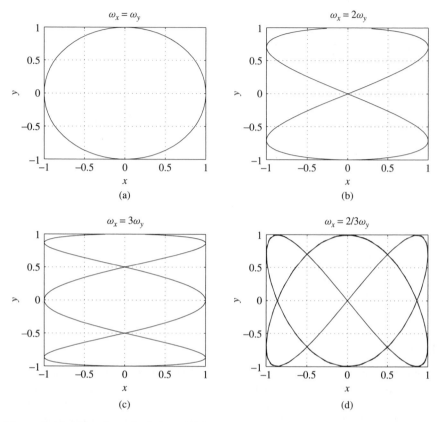

Figure 2.10   Lissajous loci for different combinations of the frequency of the orthogonal movements, ω_x/ω_y.

is given by

$$\omega_y = n_y\omega_0 \tag{2.10}$$

$$\omega_x = n_x\omega_0 \tag{2.11}$$

where n_y and n_x are integers.

For the particular situation in which $n_x = n_y$, and $\varphi_x - \varphi_y = \frac{\pi}{2}$, the locus is an ellipse (see Figure 2.10a). This is the situation usually found in resonant ultrasonic motors. The cases of different combinations of frequencies are also shown in Figure 2.10b, c and d.

The practical application to piezoelectric resonant drives involves either the combination of two orthogonal vibration modes at the same frequency or the use of what is known as the *mode conversion principle* (see Section 2.3.2).

Resonant-type piezoelectric motors can be classified into *linear* or *rotational* motors, depending on the resulting macroscopic movement. In either case, the *travelling waves* are normally used as a means of exciting the Lissajous locus at the stator–rotor interface. The next section deals specifically with rotational resonant piezoelectric motors, and in particular with *travelling wave ultrasonic motors, TWUMs*.

2.3.2 Rotational ultrasonic motors

There is no clear consensus as to who conducted the original research on rotational ultrasonic motors, but it was most likely in the Soviet Union in the early 1960s. As early as 1964, *Lavrinenko* introduced a rotational motor based on a piezoelectric ceramic converter. In the early 1970s, a number of implementations of rotational ultrasonic motors based on resonant piezoceramics were known (for instance, Siemens and Matsushita Electric Industries).

The next steps towards the development of high-performance ultrasonic motors were taken in Japan. In 1980, *Toshiiku Sashida* presented an ultrasonic motor with practical industrial applications (driving frequency 27.8 kHz, output torque 0.25 Nm, input electrical power 90 W and output mechanical power 50 W). Initial practical problems with wear at the rotor–stator interface were overcome two years later when he introduced the travelling wave ultrasonic motor, TWUM.

There are various classifications for rotational resonant piezoelectric motors. All classifications share the concept of the resulting rotational macroscopic motion and the use of at least one vibration mode excited in resonance.

According to the definition of actuators as two-port transducers, rotational piezoelectric motors have an electrical input port and a mechanical output port. The input port can be configured as a single-phase electrical port, as a two-phase port or as a multiple-phase input port. According to this criterion, a first classification of rotational piezoelectric motors relates to the number of electrical phases at the input port.

A more useful classification of resonant piezoelectric motors takes into consideration the vibration modes used to produce the elliptic microscopic motion. Generally speaking, a combination of the following resonance modes can be used to excite an elliptic motion at the rotor–stator interface:

- *Flexural Vibration Mode.* The stator of a piezoelectric resonant motor can be considered a laminate elastic structure. The laminate structure is composed of an elastic metal substrate and an exciting piezoelectric ceramic. This disk-shaped laminate structure is usually thin, in the sense that the diameter of the disk is much larger than its thickness, and it therefore meets the *Kirchhoff*'s assumption.

  For structures of this type, the Kirchhoff's assumption established that when a flexural vibration mode is excited, the planes orthogonal to the neutral plane remain orthogonal after the flexural strain is excited. In this type of vibration mode, the axial displacement is a function of the circumferential

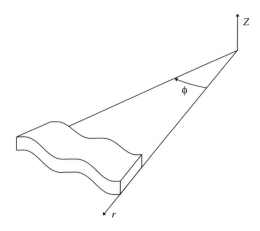

Figure 2.11   Flexural vibration mode characterized by pure bending deformations in the circumferential direction.

variable, ϕ, and the radial variable (see Figure 2.11).

$$u(\phi, r) = A \cos(k\phi) F(r) \qquad (2.12)$$

In this equation, k defines the mode order. The higher the value of k, the larger is the number of nodal diameters. In Equation 2.12, $F(r)$ is a function of the radial variable. The particular shape of $F(r)$ is also a function of the mode order; here again, the higher the mode order, the higher is the number of nodal circumferences.

- *Radial Vibration Mode.* In this type of vibration mode, the stator displacement is predominantly radial. This vibration mode is depicted graphically in Figure 2.12. Radial modes are most often used in combination with flexural modes to achieve microscopic elliptic motion.

- *Longitudinal Vibration Mode.* The predominant component in the longitudinal vibration mode is an axial displacement. Figure 2.13 shows this vibration

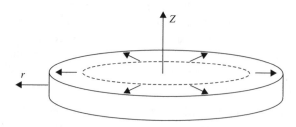

Figure 2.12   Radial vibration mode characterized by pure radial deformations along r axis.

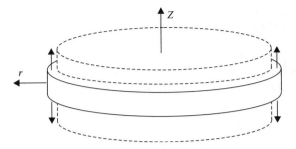

Figure 2.13 Longitudinal vibration mode characterized by normal deformations along z axis.

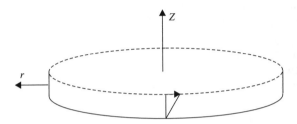

Figure 2.14 Torsional vibration mode characterized by shear deformations of the upper part of the disk with respect to the lower part along the θ axis.

mode schematically. Longitudinal vibration modes are commonly used either in combination with torsional and flexural modes or in what is known as the mode conversion approach.

- *Torsional Vibration Mode.* A torsional vibration mode is predominantly a shear strain resonance mode (see Figure 2.14). As in the previous case, torsional modes are usually combined with other orthogonal modes to obtain the required Lissajous loci.

There is a further classification of rotational piezoelectric motors based on the way the different modes are combined to achieve elliptic motion at the stator–rotor interface.

1. *Motors using resonance modes of the same type.* We know that because of the axial symmetry of disk type piezoelectric motors there are two degenerate modal shapes for each out-of-plane vibration mode. These degenerate modal shapes are shifted by a magnitude equal to $\lambda/4$ in the circumferential direction, where λ is the circumferential wavelength.

Flexural vibration modes as described above are typical examples of this situation. For this type of vibration mode, it is relatively easy to excite both degenerate modal shapes at the same resonance frequency. The well-known *travelling wave ultrasonic motors* belong to this category.

2. *Motors using resonance modes of different types at the same frequency.* In piezoelectric motors, it is possible to excite an elliptic motion at the rotor–stator interface by a combination of two vibration modes of different types.

Since different vibration modes will generally occur at different resonance frequencies, the geometry of the stator in this type of motors is designed in such a way that both modes are brought together at the same resonance frequency.

This is true of a two-piezoelectric ceramic stator in which the first longitudinal and the first transversal vibration modes are used (see Figure 2.15a). In such a stator, the frequency of these two vibration modes can be made to coincide by a proper selection of the mass' length, L_0. Under this condition, both vibration modes are excited and the corresponding Lissajous locus is achieved for the same driving frequency.

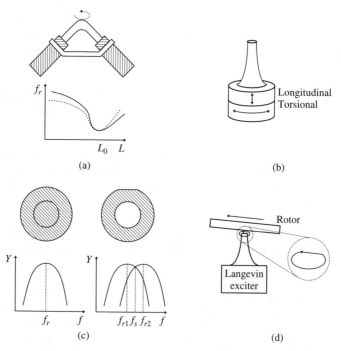

Figure 2.15 Motors exploiting different combinations of vibration modes: (a) different modes at the same resonance frequency, (b) modes of different types at different frequencies, (c) mode separation and (d) mode conversion.

3. *Motors using resonance modes of different types at different frequencies.* An approach similar to the above is when two orthogonal vibration modes are combined but one of them is excited out of resonance. As mentioned before, different vibration modes generally occur at different frequencies.

In this type of motor, two electrically independent piezoelectric ceramics are used to drive both orthogonal vibrations. Both ceramics are excited at the same driving frequency, but owing to the separation of modes, only one of them is driven in resonance. Since the generation of Lissajous loci requires the excitation of orthogonal vibration with a phase lag of $90°$, the driving electrical signals for both vibration modes are used in quadrature. Figure 2.15b shows the steps in the driving sequence for this type of motor.

4. *Motors using mode separation.* As discussed earlier, two degenerate modal shapes coexist at the same frequency in out-of-plane vibration modes of structures exhibiting axial symmetry. In such cases, a small geometrical perturbation of the structure is sufficient to produce nonsymmetric geometry, and this in turn will separate the two degenerate modal shapes.

In this situation, the use of a frequency of the driving electrical signal within the range bounded by the new resonance peaks corresponding to the separate degenerate modes would result in combined excitation of the two modes (see Figure 2.15c).

5. *Motors using mode conversion.* In this type of motor, the stator is excited in a single longitudinal vibration mode. The longitudinal vibrator is a *Langevin* transducer (Ueha *et al.* (1993)), which is placed nearly perpendicular ($\approx 85°$) to a disk rotor.

The longitudinal displacement of the Langevin vibration produces an impact, and, thus, the longitudinal displacement is converted to a transversal displacement (see Figure 2.15d). Some of the limiting factors in this kind of motor are the heavy wear at the interface and the difficulty in reversing the rotation direction.

Travelling wave ultrasonic motors

One of the most successful implementations among the various different resonant-type rotational piezoelectric motors has been what is known as the travelling wave ultrasonic motor, TWUM. The term *ultrasonic* refers to the fact that the driving frequency used to excite these motors lies in a frequency range above 20 kHz and is, hence, well beyond the limits of human hearing.

TWUMs make use of two degenerate flexural vibration modes at the same frequency to achieve elliptic motion at the rotor–stator interface. Therefore, they belong to the first type of motors as described above (see Figure 2.16).

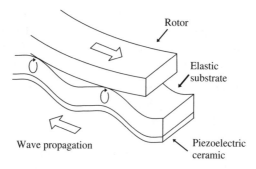

Figure 2.16   Schematic view of the travelling wave motor concept.

A *travelling wave* is generally described by a mathematical function having the form:

$$u(\phi, t) = Af(k\phi \pm wt) \qquad (2.13)$$

In rotational motors, ϕ is the circumferential variable, k defines the mode order, w is the frequency of the driving electrical signal and t is the time. In Equation 2.13, the minus sign leads to a wave travelling in the direction of increasing ϕ, while the positive sign leads to a wave travelling in the negative direction of ϕ.

The mathematical expression of a travelling wave can be expanded as a sum of two *standing waves*. By assuming a sinusoidal function for the travelling wave and taking into account that the axial displacement, $R(r)$, is a function of the radial distance, Equation 2.13 can be written as

$$u(\phi, r, t) = AR(r)\cos(k\phi)\cos(wt) \mp AR(r)\sin(k\phi)\sin(wt) \qquad (2.14)$$

where the function $R(r)$ describes the axial displacement of the stator as a function of the radial variable r.

Equation 2.14 suggests how a travelling wave can be excited in the ultrasonic motor stator. At a first glance, Equation 2.14 will show the presence of a *cosine*, and a *sine* geometrical pattern. The wavelength of these patterns, λ, is defined by k, the mode order, so that $\lambda = 2\pi/k$.

It is common practice to fix both geometrical patterns in the stator's structure during poling of the piezoelectric ceramic used to drive the travelling wave. In practice, this is done by an alternating poling pattern in the direction of the thickness of the piezoelectric ceramic. Usually, half of the piezoelectric disk is used to define what is called the *sine mode* while the other half is used for the *cosine mode*. Each mode is obtained by alternating the poling direction in adjacent circumferential angular sectors of an arc of $\lambda/2$. Both modes are circumferentially shifted apart by a magnitude of $\pi/2k$, which corresponds to $\lambda/4$. The poling pattern is described schematically in Figure 2.17a.

Once the piezoelectric ceramic is poled according to the *sine* and *cosine* modes, a two-phase electrical excitation circuit is configured. In this way, one of the sides

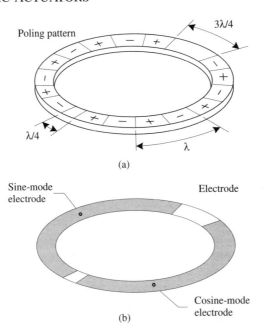

(a)

(b)

Figure 2.17   Configuration of the excitation of a travelling wave. (a) Process for poling *sine* and *cosine* modes in the piezoelectric ceramic and (b) use of independent electrodes for both modes.

of the piezoelectric ceramic becomes the reference electrode. Independent electrodes are used on the other side of the ceramic for both modes (see Figure 2.17b).

By selecting the appropriate phase of the electrical signals applied to both phases, it is possible to select the turning direction of the travelling wave, and, thus, the direction of rotation of the TWUM.

Let us analyze the case of a thin laminate structure in which a travelling wave is excited (see Figure 2.18). As noted earlier, Kirchhoff's assumption for flexural vibration of thin structures posits a motion in which planes perpendicular to the neutral plane remain perpendicular after bending. In Kirchhoff's theory, points on the neutral plane exhibit pure axial displacement in flexion. According to this assumption, the displacement of a point Q at the rotor–stator interface is described by the following expression:

$$\vec{u}_Q = u(r, \varphi, t)\vec{e}_z - a u_{,r}(r, \varphi, t)\vec{e}_r - (a/r)u_{,\varphi}(r, \varphi, t)\vec{e}_\varphi \qquad (2.15)$$

where $u_{,r}(r, \varphi, t) = \frac{\partial u}{\partial r}$, $u_{,\varphi}(r, \varphi, t) = \frac{\partial u}{\partial \varphi}$, $\vec{e}_z$, $\vec{e}_r$ and $\vec{e}_\varphi$ are unit vectors in axial, radial and circumferential directions respectively and a is the distance from the rotor–stator interface to the neutral plane.

It can readily be shown that the projection of the trajectory of point Q described by Equation 2.15 onto a tangential plane is an ellipse. It can further be demonstrated

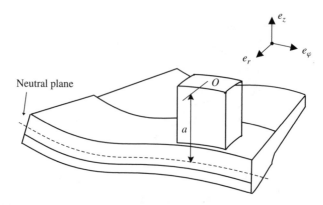

Figure 2.18   Schematic representation of a toothed TWUM stator.

that the velocity of point Q in the upper part of the trajectory has a horizontal
component only.

In TWUMs, both rotor and stator are prestressed against each other by a compressive force. The tangential motion of point Q in the upper part of the trajectory
is then transmitted through friction to the rotor. The rotor's angular velocity can
be estimated on the assumption that there is no sliding at the interface. The linear
velocity of the stator's contact point Q is

$$v_t = -\Omega \frac{a}{r_B} k R(r_B) \tag{2.16}$$

where Ω is the rotational frequency of point Q while describing the elliptic trajectory depicted in Figure 2.19, r_B is the radius at the point of contact between stator
and rotor and $R(r_B)$ is the axial displacement at radius r_B.

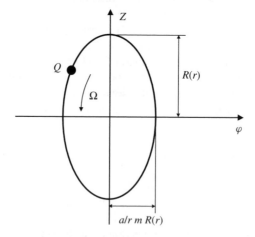

Figure 2.19   Idealized trajectory of a point Q on the stator–rotor interface.

Assuming there is no sliding, the angular velocity of the rotor becomes

$$\dot{\Phi}_{Rotor} = -\Omega \frac{a}{r_B} m \frac{R(r_B)}{r_B} \qquad (2.17)$$

A first glance shows that some qualitative conclusions can be drawn from the analysis of Equation 2.17:

1. The equation suggests a high transmission (reduction) ratio between the driving frequency Ω and the motor's angular velocity $\dot{\Phi}_{Rotor}$.

   In fact, since the disk is thin, the ratio $\frac{a}{r_B}$ is small, typically of the order of 0.01. However, the ratio $\frac{R(r_B)}{r_B}$ is even smaller, since $R(r_B)$ is of the order of a few microns. The combination of these reduction ratios leads to an aggregate reduction of approximately 1:50,000.

   The TWUM can therefore be said to have a *large intrinsic transmission ratio*. This in turn means that motor operation is characterized by a low rotational speed (a few hundred rotations per minute) and a high torque.

2. Equation 2.17 suggests various different ways in which the TWUMs can be controlled. It can readily be appreciated that the motor's angular velocity is proportional to the driving electrical frequency, Ω, and to the amplitude of the axial vibration at the contact point $R(r_B)$. This suggests that both the amplitude and the frequency of the driving electrical signal could be used to control the angular velocity of the TWUMs.

   Even though the above consideration is correct, care has to be taken since operation in the proximity of resonance or antiresonance peaks leads to hidden secondary effects that become dominant when the driving frequency is selected as the input control variable. In fact, as noted in Section 2.3, the frequency operation range for TWUMs is commonly the region of smooth impedance decay immediately above the antiresonance peak.

   In this region, a low frequency means proximity to the antiresonance peak, and this in turn means high amplification of the mechanical vibration, $R(r_B)$. The latter is dominant, and the motor's angular velocity is maximum at low frequency. This is the opposite of what one might expect from a first glance at equation 2.17.

2.3.3 Linear ultrasonic motors

The previous section contained an analysis of rotational resonant piezoelectric drives and classified them according to the particular vibration modes that are used to excite elliptic motion at the rotor–stator interface. This was followed by a similar analysis in terms of the way these vibration modes are combined to generate driving elliptic motion.

Linear resonant piezoelectric drives can be analyzed in a similar way. In order to avoid unnecessary duplication in the analysis of resonant drives, this is restricted to the case of *travelling wave linear motors, TWLMs*.

Travelling wave linear motors, TWLMs

TWLMs have specific particularities whose analysis is relevant at this stage. In the case of rotational travelling waves, the elastic substrate supporting the travelling wave is an infinite continuum. In the particular case of linear travelling wave motors, the elastic substrate is *per se* finite.

The governing equation for the transverse motion of a beam or rod can be obtained from the statement of equilibrium conditions for a small element of the rod (Graff (1975)):

$$\frac{\partial^4 v}{\partial x^4} + \frac{\rho A}{EI} \frac{\partial^2 v}{\partial t^2} = q(x, t) \tag{2.18}$$

where v is the transversal displacement of the rod points, x is the longitudinal variable describing the linear position in the rod, ρ is the density of the elastic substrate, A is the section area, E is the Young's modulus of the substrate, I is the moment of inertia of the cross section, and $q(x, t)$ is a distributed load on the rod in the direction of y. If we assume that no load is being applied to the rod, Equation 2.18 becomes

$$\frac{\partial^4 v}{\partial x^4} + \frac{\rho A}{EI} \frac{\partial^2 v}{\partial t^2} = 0 \tag{2.19}$$

Equation 2.19 can be solved for the transversal displacement $v(x, t)$:

$$v(x, t) = a \sin(kx - wt) \tag{2.20}$$

Equation 2.20 represents a sinusoidal wave travelling in the positive direction of x. The longitudinal displacement of the rod, $u(x, y, t)$, can be derived by means of the following expression:

$$u(x, y, t) = -y \frac{\partial v}{\partial x} = -ayk \cos(kx - wt) \tag{2.21}$$

The linear combination of the transverse and longitudinal displacement of Equations 2.20 and 2.21 results in an elliptic motion of the points in the rod (conforming to the Lissajous loci described in previous sections).

In finite beams or rods, the travelling wave, $v(x, t)$, will be subject to a reflection process at the free end of the elastic domain. As a consequence of this process, a travelling wave is also generated in the opposite direction, $v_o(x, t)$. The mathematical expression for the latter is

$$v_o(x, t) = a \sin(kx + wt) \tag{2.22}$$

The linear combination of the travelling waves represented by Equations 2.20 and 2.22 results in a standing wave, $u_s(x, t)$, in the beam or rod:

$$v_s(x, t) = a \cos(kx) \cos(wt) \tag{2.23}$$

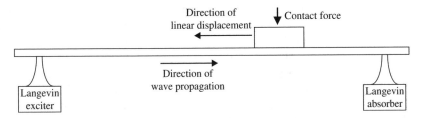

Figure 2.20   Schematic representation of a travelling wave linear motor.

The longitudinal displacement corresponding to this situation is

$$u_s(x, y, t) = -y\frac{\partial v_s}{\partial x} = -ayk\sin(kx)\cos(wt) \tag{2.24}$$

For a particular point (x_1, y_1) in the rod elastic domain, the expression for the transversal and longitudinal displacement in the rod reduces to

$$u_s(t) = K_u\cos(wt)$$
$$v_s(t) = K_v\cos(wt) \tag{2.25}$$

The combination of u_s and v_s according to Equation 2.25 results in a linear displacement of the points in the elastic domain. No elliptic motion is established as a consequence; friction transmission to a linear slider cannot be realized using this scheme.

What this analysis indicates in practice is the need for a mechanism of absorption of the travelling wave at the opposite end of the finite elastic domain. In common practice, Langevin vibrators are used at either end of the elastic domain (see Figure 2.20). One of the Langevin vibrators is used to excite a wave travelling in the direction of the other vibrator. The latter then acts as a vibration absorber. If the situation is reversed, the wave travels in the opposite direction, as does the linear slider.

2.4   Nonresonant piezoelectric actuators

2.4.1   Bimorph actuators

The word "bimorph" is a registered trademark of Morgan Electro Ceramic, but it has come to be the commonly adopted name for bending piezoelements comprising two bonded ceramic plates. Bending piezoelectric elements comprising a multiplicity of bonded ceramic plates are called *multimorphs*. The particular case of a single piezoelectric element bonded to an elastic substrate is usually referred to as a *unimorph*.

Bimorphs are thin bending piezoelectric elements made up of a combination of two very thin piezoceramic plates bonded together to a thin metal beam. The poling directions of the ceramic plates in the bimorph and the selection of electrodes are so configured that when electrically powered, one of the ceramics will expand while the other contracts. The combined mechanical interaction between the two ceramics causes bending of the bimorph.

Also, when the bimorph structure is bent, a voltage is developed across the electrodes, as one might expect given the reversibility of the piezoelectric effect.

Piezoelectric bimorphs can be poled and configured to operate in series or in parallel. See Figure 2.21 for a schematic representation of both configurations. The parallel configuration has the ceramic plates poled in opposite directions, while the series configuration has them poled in the same direction.

In the electrical connections for the parallel configuration, one electrode is connected to the central metal beam, while the second electrode is connected to both outer sides of the ceramics. In the series configuration, each electrode is connected to each outer side of the ceramics and no electrical connection is established with the central metal vane.

As to the applicability of bimorphs as actuators, the parallel configuration gives twice as much deflection as the series configuration. This follows from the fact that the full driving voltage is applied to each ceramic plate.

The intrinsic characteristics of bimorphs as actuators are

1. High deflection and low force with a relatively low driving voltage (as compared to direct piezoelectric extenders).

2. Voltage-limited operation for both DC and AC applications of the driving voltage. This is because in both series and parallel configurations, one of the ceramic plates is always subject to an electric field opposite to the original polarization.

3. Stress-limited operation in resonance, resulting in driving voltages about one order of magnitude less than for nonresonance driving.

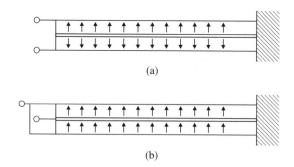

Figure 2.21   Schematic representation of the series and parallel bimorph configuration.

2.4.2   Stack piezoelectric actuators

Stack piezoelectric actuators are made of thin piezoelectric laminae stacked in such a way that a common electrode is deposited between two consecutive laminae. After this stacking process, all electrodes in alternating positions are electrically connected, so that there are two electrical terminals. See Figure 2.22 for a schematic representation of the stacking process.

Each lamina in the stack piezoelectric structure is a few tens of micrometers thick. All the laminae are poled in the direction of the thickness. In addition, the laminae are stacked in such a way that consecutive laminae have opposite poling directions. The result of the electrode disposition described above and the process of alternating lamina poling directions is a device that acts electrically in parallel and mechanically in series. One direct consequence of this is that a high mechanical displacement is achieved for a low applied electric field.

If a voltage V is applied to the stack piezoelectric actuator, every single lamina will respond according to the following equation:

$$\Delta l_i = V d_{33} \tag{2.26}$$

where d_{33} is the piezoelectric coefficient and Δl_i is the mechanical displacement for the lamina.

The total displacement for the particular configuration of the piezoelectric stack is

$$\Delta l = \sum_{i=1}^{n} \Delta l_i = n V d_{33} \tag{2.27}$$

In the event that higher range displacements are required, piezoelectric stacks can be combined in turn. Piezoelectric stacks are commercial devices. Table 2.2 shows a comparative list of several commercial piezoelectric stacks, with particular reference to the operational data for these devices. The force level is comparatively

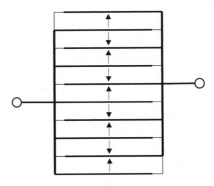

Figure 2.22   Schematic representation of the stacking and electrode configuration.

Table 2.2   Comparative technical features of commercial piezoelectric stacks

	Matroc 7111-03031	Matroc 7111-05051	Ferroperm Pz27-40x033	Ferroperm Pz29-27x067
Dimensions, mm	$3.5 \times 3.5 \times 3.5$	$5 \times 5 \times 3.5$	$6.8 \times 4.7 \times 1.7$	$6.7 \times 4.6 \times 2.2$
Capacitance, nF	70	120	–	–
Voltage, V	$+150--20$	$+150--20$	100	200
Maximum stroke, μm	2	2.8	1.6	3.3
Maximum force, N	425	850	–	–
Maximum Temperature, C	<75	<75	<250	<150

higher as compared to piezoelectric unimorphs or bimorphs, while the available displacements are of the order of a few microns.

The use of stacked piezoelectric actuators can produce two important consequences. First, because of the low stroke availability, the application of stacks may require the use of mechanical transmission stages for increased displacement; secondly, the positioning accuracy of piezoelectric stacks can be very high, of the order of nanometers. Because of their operational characteristics, these actuators are normally used in micropositioning stages of vibration suppression systems.

Piezoelectric stacks belong to the nonresonant category of piezoelectric actuators. This means, in practice, that they can be driven in a frequency range starting from static application of driving voltages. This kind of actuator cannot take advantage of intrinsic mechanical amplification at resonance, and, therefore, high displacements require high electric fields.

As noted earlier, in the process of poling, the material of piezoelectric ceramics is cooled down from a high temperature while the poling electric field is being applied. Under operational conditions, there is a threshold for the applicable driving voltage. This threshold is imposed to prevent depolarization of the piezoelectric ceramic as a result of high voltages being applied in the opposite direction to the poling voltage. In order to minimize the chance of depolarization of the ceramic, an offset is usually applied to the driving voltage to ensure that there is no depolarization.

When piezoelectric actuators are driven at alternating voltages, there is an increase in the material temperature, that is, part of the reactive power required to drive a capacitive electrical load is converted to heat. This has two main drawbacks: first, the temperature rise must be limited so that the material does not reach its Curie temperature, T_c (to prevent depolarization); second, there may be alteration of the adhesive layers between consecutive ceramic laminae.

The time response of piezoelectric actuators is highly dependent on the electrical capacitance of the device, which is usually large. The charge time for the stack actuators is limited by the power source being used, but the discharging time will mainly depend on the actuator capacitance, and this will impose a very strict limit on the response time. Piezoelectric stack manufacturers usually provide maximum charge and discharge electrical currents in their data sheets.

2.4.3 Inchworm actuators

The inspiration for the conceptual operating principle of inchworm piezoelectric actuators is biological, specifically the movement of some earthworms. Piezoelectric inchworm actuators are characterized by the long strokes of which they are capable; these are generally limited by the length of the rotor.

The linear inchworm actuator comprises three independently driven piezoelectric ceramics. The three piezoelectric ceramics can be configured for a variety of implementations, but the most typical one is depicted in Figure 2.23. In this configuration, ceramics 1 and 3 are used in a radial actuation mode to clamp the rotor, while ceramic 2 is driven in an axial mode to effectively produce the displacement.

The axial displacement of the rotor is then transformed into a cycle:

1. The first ceramic (1) is driven so that it deforms radially in order to clamp the rotor.

2. The second piezoelectric ceramic (2) is actuated in axial mode. If the mid position of the second piezoelectric ceramic is taken as the reference position for the axial displacement of the rotor, this step leads to an axial displacement that is half the displacement of the second piezoelectric ceramic.

3. The third ceramic (3) is then actuated in a manner similar to the first one. It clamps the rotor while the second ceramic is still actuated.

4. At this stage, the first ceramic (1) is relaxed so that it releases the rotor.

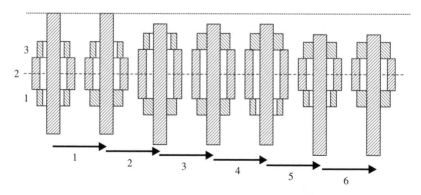

Figure 2.23 Schematic representation of the stepping process in inchworm piezoelectric motors.

5. The second ceramic (2) is now relaxed and as a result the rotor is again displaced axially, by half the total relaxation of the second piezoelectric ceramic.

6. This last step completes the cycle as the third ceramic (3) is relaxed.

If this operation sequence is repeated, the resulting operational behavior is similar to that of "stepping" motors. The rate of axial displacement will range from a few nanometers per second to a few meters per second, depending on the duration of the entire cycle and the voltage applied.

2.5 Control aspects of piezoelectric motors

2.5.1 Control circuits and resonant drivers

It is common practice in textbooks on mechatronics to show sensors, actuators and controllers as constitutive components in control motion systems. This was already stressed in Chapter 1 when dealing with the role of actuators in mechatronic systems.

As noted earlier, an actuator can in itself be considered a mechatronic system. In the particular case of resonant-type piezoelectric motors, this is apparent from the analysis of their operation conditions. During normal operation of a resonant piezoelectric motor, the driving voltage, the system temperature or the actuator load will be subject to variations. Consequently, the resonant characteristics of the system will be altered. This is commonly sufficient to cause alterations in the operation point.

Optimum tracking of the operation point is therefore required to overcome temperature, voltage or load changes. The adaptation of driving conditions can be optimized for tracking the operation point following a threefold approach:

1. *Selection of efficient driving signals.* Because piezoelectric motors are reso- nant, a wise selection of driving signals would aim at providing pure tones or at least at shifting harmonics apart.

2. *Signal amplification and tuning.* Switching signals used to drive capacitive loads usually cause unacceptable current spikes. This can only be over- come by proper design of a resonant driving circuit suitably attuned to the mechanical resonance characteristics of the piezoelectric motor.

3. *Resonance tracking.* The shift in the mechanical resonance characteristics of the piezoelectric motor must be tracked to ensure an optimum operation point.

In the coming paragraphs, we focus on an analysis of the three different steps for optimum operation of resonant piezoelectric motors. The three-step approach is a typical example of a mechatronic approach to system design: *the analysis of the mechanical characteristics of the plant (piezoelectric motor) leads to the formulation of control strategies (which are based on tuning electronic drivers) to track the operation point and thus enhance overall operation and performance.*

Selection of efficient driving signals

The piezoelectric motor is a resonant structure that behaves as a mechanical filter. The optimum driving signal for a piezoelectric ceramic is a pure sinusoidal signal tuned to the resonance frequency of the mechanical part. Any undesired harmonic in the driving signal will be filtered by the piezoelectric ceramic, causing overall heating and loss of efficiency.

Digital electronic circuits are preferred to their analog counterparts. The solution is usually to use switching techniques to set up the driving signal. Switched signals can be configured to reduce the undesirable effects of harmonics of the fundamental resonance frequency.

According to Schaaf and van der Broeck (1995), if, for instance, bipolar symmetrical voltage pulses are used in particular, the frequency spectrum can be demonstrated to be

$$V_n = \frac{4}{\pi} V \frac{\sin n\pi w}{n} \frac{[1 - (-1)^n]}{2} \tag{2.28}$$

where V is the DC supply voltage, n is the number of the harmonic and w is the duty cycle of the signal.

Selection of a bipolar symmetrical signal results in zero even harmonics. Furthermore, if the duty cycle of the driving signal is $1/3$, the third harmonic vanishes (see Figure 2.24). The combination of these two conditions produces an acceptable driving signal that is practically equivalent to a pure tone at the resonance frequency.

Signal amplification and tuning

The inverter output voltage of the previous stage cannot be applied directly to the motor since the switched voltage would cause high current spikes at the clamped capacitance C_0.

In order to overcome this problem, an inductor has to be placed in series to the motor. The inductor will cause a slight drop in the fundamental voltage. This must be limited by means of an additional serial capacitor. The inductor and the capacitor together form a serial resonant filter (see Schaaf and van der Broeck (1995)) whose resonant frequency must be tuned to that of the motional impedance of the motor.

The inverter voltage is adapted to the required voltage at the motor by means of an inductive transformer. In an optimum situation, the leakage inductance of the transformer can be used as the reactive component of the resonant series filter (see Schaaf and van der Broeck (1995)). The transformer's secondary inductance is used to compensate for the clamped capacitance of the motor. Again, the parallel resonance frequency must be tuned to the resonant frequency of the motional

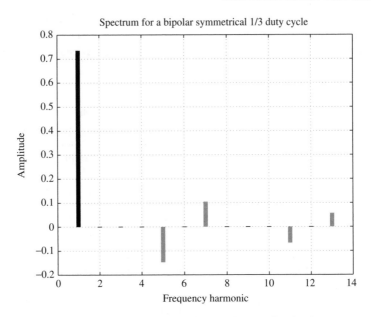

Figure 2.24   Amplitude of the various different harmonics in the spectrum of a switched bipolar symmetrical signal with a duty cycle of 1/3.

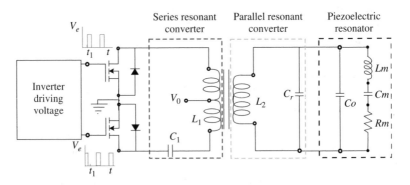

Figure 2.25   Schematic representation of the signal amplification and tuning circuit. Proper selection of the various different electric parameters produces resonance matching and parasitic resonance rejection.

impedance of the motor. The design of the resonant electrical circuit must fulfill the condition of matching the resonant frequency of the motional impedance, the series resonant filter and the parallel resonant filter.

Following the nomenclature of Figure 2.25, the condition for matching the series and parallel filters to the motional impedance of the piezoelectric motor is formulated mathematically by the following expression:

$$L_m C_m = L_s C_1 = L_2(C_0 + C_2) \tag{2.29}$$

In addition, care must be taken to avoid excessive reactive currents due to parasitic resonance in the circuit. The first parasitic resonance, w_{p1}, (involving the path comprising C_1 and L_1) affects transient situations when switching on because its resonance frequency is relatively lower. The second parasitic resonance, w_{p2}, is due to the serial connection of clamped capacitance, compensation capacitors and the leakage inductance. Since this second parasitic resonance frequency is relatively higher than the motional resonance frequency, care must be taken not to let this parasitic component come into proximity with harmonics of the fundamental frequency.

Here, we can profitably select bipolar symmetrical switched signals with a duty cycle of $1/3$. For this particular signal, the first nonzero harmonic is $5w_r$. It would be sufficient to limit this second parasitic resonance, for instance, to $w_{p2} \le 3w_r$.

The effect of unmatched power drives is twofold: first, unadapted impedances produce a decrease in the driving voltage; second, they also result in a phase shift with respect to the reference phase shift. This can be seen particularly in Figure 2.26. Figure 2.26 shows the driving voltage signals for a two-phase TWUM. Ideally, both signals should exhibit a phase lag of 90° and similar amplitudes when

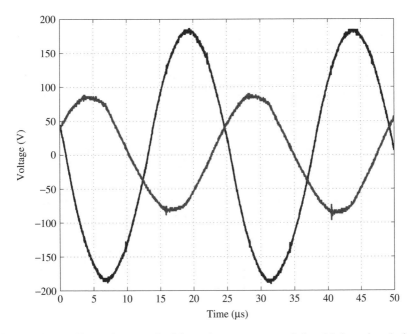

Figure 2.26   Effect of unmatched impedance on one of the driving signals in a two-phase TWUM.

perfect impedance matching is achieved. In this case, the first phase is impedance matched but phase two is unmatched. Consequently, the voltage level for the second driving signal is reduced and the phase lag is shifted so that both signals are nearly out of phase.

Resonance tracking

As explained in previous sections, a piezoelectric resonator is characterized by a capacitive electrical load in a frequency range below the resonance frequency and above the antiresonance frequency. In between, the electrical equivalent to the piezoelectric ceramic is inductive. This means that the piezoelectric ceramic becomes a pure resistive electrical load at resonance and antiresonance. This is consistent with the fact that at resonance and antiresonance the reactive part of the electrical impedance vanishes, thus producing peak efficiency.

The resonance and antiresonance frequency of a piezoelectric actuator is generally subject to perturbations during operation. For optimal operation, a tracking electronic drive is required.

The functional characteristics of such a tracking drive are depicted in the block diagram of Figure 2.27. The blocks in the tracking system are as follows:

1. *Error phase detector.* The role of this block is to provide an error signal proportional to the phase error between driving voltage and current drawn.

2. *Loop controller.* The loop controller receives the phase error between current and voltage as an input and provides a control signal that asymptotically tracks the resonance frequency of the piezoelectric actuator.

3. *The plant.* This block represents the piezoelectric actuator itself. It receives the control action and performs at resonance independently of external perturbations.

In practical terms, the design of a resonance frequency–tracking electronic drive is commonly based on a *phase-locked loop*, PLL, technique. A PLL consists of two main building blocks, namely the *phase detector*, PD, and the *voltage controlled oscillator*, VCO.

In a common implementation, a phase detector works as an up–down counter, in which the up count is edge triggered by the first input signal (the current; see

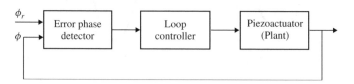

Figure 2.27 Functional block diagram of the tracking loop for resonance and antiresonance operation.

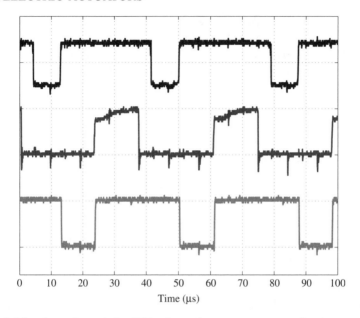

Figure 2.28 Operation of the PLL phase detector as an up–down counter edge triggered by two input signals. The upper line represents zero crossing points of the current drawn; the middle line is the switching driving voltage; and the lower line is the switching output signal from the phase detector.

the upper line in Figure 2.28) and the down count is also edge triggered by the second input signal (the voltage; see the middle line in Figure 2.28).

According to this description, the phase detector output is made up of a train of pulses, whose width is proportional to the phase lag between voltage and current drawn (see the lower line in Figure 2.28).

The phase detector provides an output signal proportional to the difference in phase between current drawn and voltage:

$$v_e = K_d(\theta_i - \theta_o) = K_d\varepsilon \qquad (2.30)$$

$$\frac{V_e(s)}{\varepsilon(s)} = K_d \qquad (2.31)$$

where K_d is the phase detector gain.

In order to obtain the desired performance from the error phase detector in Figure 2.27, a low pass filter can be added to the output of the PLL phase detector. The combination of the phase detector and the low pass filter will produce an analog error signal proportional to the phase difference between voltage and current.

The VCO generates a switching signal with a fundamental frequency proportional to its input voltage. The duty cycle of the switching signal can generally be selected to suit the requirements of the application and will not interfere with the phase detector described above. As explained at the beginning of this section, a suitable choice for the duty cycle would be around 1/3.

The VCO will provide a deviation from the central frequency, which is proportional to the analog error signal from the phase detector and loop filter, v_f. This is mathematically expressed as

$$\Delta\omega = K_0 v_f \tag{2.32}$$

Since frequency is a derivative of phase, the above formulation for the functional characteristics of the VCO can be described as

$$s\Theta_{vco}(s) = \frac{K_0}{V_p(s)} \tag{2.33}$$

It can readily be appreciated that Equation 2.33 describes the VCO as a pure integrator. As a result, the closed loop transfer function of the tracking loop is always a Type I loop.

If the tracking circuit implements an appropriate controller together with the VCO in a cascade configuration, this will give the functionality of the loop controller in Figure 2.27. To be able to reject permanent DC components in the error signal, the appropriate low pass filter must be selected.

In the simplest configuration, a classic passive low pass filter can be implemented. As discussed by Gardner (1979), a passive low pass filter gives good results in most tracking applications. The transfer function for the low pass filter is

$$\frac{V_f(s)}{V_e(s)} = \frac{1 + s\tau_2}{1 + s\tau_1} \tag{2.34}$$

The open loop transfer function for the tracking circuit can be written as

$$G(s) = K_d \frac{K_0}{s} \frac{1 + s\tau_2}{1 + s\tau_1} \tag{2.35}$$

The closed loop transfer function corresponding to the tracking circuit comprising the phase detector (Equation 2.31), the loop low pass filter (Equation 2.34) and the VCO (Equation 2.33) is

$$H(s) = \frac{G(s)}{1 + G(s)H(s)} = \frac{K_d K_o / \tau_1 (1 + \tau_2 s)}{s^2 + s(1 + K_o K_d \tau_2)/\tau_1 + K_o K_d / \tau_1} \tag{2.36}$$

The stability of the tracking system can readily be shown using the *final value theorem*, whose mathematical formulation is

$$\lim_{t \to \infty} h(t) = \lim_{s \to 0} s H(s) \tag{2.37}$$

For the particular case of classic low pass passive filters, the system is Type I, that is, there is one perfect integrator in the tracking loop–the one provided by the VCO. When directly applied, the final value theorem predicts that the system will track step changes in the phase error without permanent phase error in the steady state.

However, the system will lead to permanent constant error in the steady state if a ramp phase error is introduced. If we wish to ensure zero tracking error upon application of ramp phase errors, the loop-tracking circuit should be completed either by a classic PI controller or by making use of active filters with a transfer function:

$$\frac{V_f(s)}{V_e(s)} = \frac{1 + s\tau_2}{s\tau_1} \tag{2.38}$$

In that case, the tracking loop would be converted to a Type II system comprising two perfect integrators. Consequently, ramp changes in the phase error would be perfectly tracked without permanent steady state errors.

2.5.2 Control of nonresonant actuators

As we discussed in previous sections, the electrical equivalent load of a piezoelectric actuator driven in a frequency band outside the resonance–antiresonance region is a capacitor. This capacitive load is larger where stack piezoelectric actuators are concerned. For a capacitive load, the following approximate electromechanical relations apply:

$$\Delta l \approx \Delta Q \tag{2.39}$$

$$v_p \approx i$$

$$a_p \approx \frac{di}{dt}$$

Hysteretic behavior of voltage-driven applications

It is known (see for instance Dörlemann *et al.* (2002)) that in open loop control of piezoelectric actuators the relationship between displacement and driving voltage is nonlinear and hysteretic. On the other hand, the relationship between displacement and charge or current drawn is quasilinear and nearly nonhysteretic. This situation is illustrated in Figure 2.29.

The effect of the nonlinear, hysteretic behavior in the voltage–strain relationship produces what is called multimode excitation of voltage-driven piezoelectric actuators. The situation can be summarized as follows: when a pure tone voltage excitation is applied to the actuator, the corresponding mechanical vibration will exhibit a nonzero spectrum at the different harmonics of the fundamental voltage frequency.

Then again, if a pure tone current signal is applied to the actuator, the resulting mechanical vibration will also be a pure tone. As a result, there is increased heating when voltage-driven actuators are used, and thus the dynamic range of the system is reduced. This situation is summarized in Figure 2.29b.

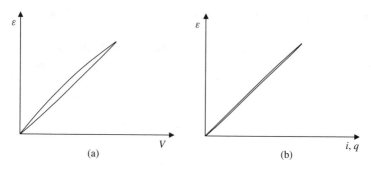

Figure 2.29   Linearization and hysteresis compensation in piezoelectric actuators by means of current or charge control.

Control of actuator stiffness

In normal, passive materials, the mechanical stiffness, K, is defined as the ratio of applied force to deformation:

$$K = \frac{dF}{dl} \tag{2.40}$$

In the case of piezoelectric actuators, because of the active nature of the material there is the possibility of *programmable stiffness*. The force-deformation ratio for piezoelectric actuators is highly dependent on the electrical boundary conditions applied to the actuator electrodes.

In this context, there are three possible situations, depending on the electrical boundary conditions:

1. *Short-circuited or voltage-controlled actuators.* In this situation, a "low" stiffness is achieved. Under deformation, the charge generated by the piezo-electric effect is free to flow and equilibrate.

2. *Open circuit or charge-/current-controlled actuators.* In this situation, the charge is blocked at the electrodes of the actuator when a force is applied. Blocking of the charge (either because the circuit is open or because the control fixes its value) results in an electric field that will oppose the force. The outcome of this situation is greater stiffness (twice as much as in case 1).

3. *Impedance-controlled drives.* In this case, the control loop applied to the piezoelectric actuator sets the reference impedance for the piezoelectric drive. At the upper limit, when the reference impedance is very high, the control strategy is equivalent to a position control of the piezoelectric motor, and the apparent stiffness of the actuator is virtually infinite.

In this case again, the charge or current control of piezoelectric drives opens up the possibility of dynamic modulation of piezoelectric drive stiffness.

Current control of piezoelectric actuators

It is clear from the above analysis that current control of piezoelectric drives offers significant advantages over voltage control:

- hysteretic behavior is reduced to a minimum,

- the displacement to drive signal relationship is linearized,

- heat losses are reduced, leading to better dynamic driving and

- the stiffness of the actuator can be modulated.

2.6   Figures of merit of piezoelectric actuators

The various different types and actuation principles of piezoelectric actuators cover a wide range of operational characteristics. They can provide short stroke operation (a few μm), both rotational and translational, with sub-nanometer resolution (piezoelectric stacks), but they can also offer unlimited stroke operation (inchworm and travelling wave concepts) in linear and rotational actuators.

The following sections analyze the main operational characteristics of piezoelectric actuators and their behavior upon scaling.

2.6.1   Operational characteristics

Static performance

The piezoelectric effect in piezoelectric ceramics leads to static strains of 0.1–0.6%, depending on the material's characteristics. These figures are as high as 1.5–1.7% in the case of single-crystal piezoelectric ceramics.

Similarly, in the case of direct application of the piezoelectric effect, blocked forces are subject to an upper pressure limit of the order of 100–110 MPa; for single crystals, however, the maximum pressure is as high as 130–140 MPa.

Direct application of the piezoelectric effect is not common. Actuators generally employ specific geometrical configurations to achieve output mechanical energy, and the available stroke and force will be highly dependent on the configuration. We would note a number of characteristics of static performance of nonresonant piezoelectric actuators:

1. *Stack piezoelectric actuators.* These provide high force and low stroke. The maximum stroke may vary in absolute terms from a few micrometers to a few millimeters. The blocked force varies within the range of 10^2–10^5 N.

2. *Piezoelectric benders.* Multimorph configurations produce a medium stroke and very limited force. In absolute terms, the stroke can vary between 10^{-2} and 10 mm and thus occupies the range between stacks and inchworm actuators (see next item). This configuration produces virtually no force, and values tend to be situated in a range of 10^{-2}–10^2 N.

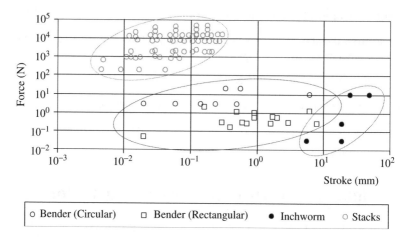

Figure 2.30   Maximum stroke versus blocked force for nonresonant piezoelectric
actuators.

3. *Inchworm actuators*. The force delivered by these actuators is in the same
range as benders, but the stroke can be very high and may even be unlimited
in some configurations. These are stepper motors and may present resolutions
down to the nanometer scale.

The relative positions of piezoelectric actuators in terms of stroke and force are
shown in Figure 2.30. They complement each other in both actuation indicators
and span a suitably wide area on the stroke versus force plane.
Resonant piezoelectric actuators are characterized by unlimited travel or stroke,
and so their static performance is best characterized in terms of maximum speed
(either rotational or translational) and stall torque or force respectively.
The data for rotational and translational actuators is scanty. There are only a
few implementations commercially available, but interesting conclusions can be
derived from a comparison of the performances of resonant piezoelectric drives
and their electromagnetic counterparts, which have become the principal actuator technology.
Travelling wave ultrasonic (rotational) motors, TWUMs, complement electromagnetic DC motors in terms of both torque and (rotational) speed. For output
mechanical power up to 7–8 W, TWUMs can be driven at a maximum speed of
the order of 1000 rpm while the stall torque can be up to 1 Nm. Compare this to
DC motors (with the same power limitation), which are usually driven at rotational
speeds in the region of 10^4 rpm with maximum stall torque of 10–100 mNm. This
situation, depicted in Figure 2.31, indicates that mechanical impedance matching
is typically better in the piezoelectric technology.

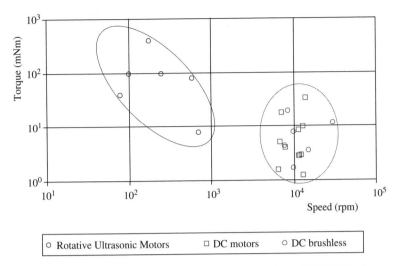

Figure 2.31 Comparison of speed–torque characteristics of TWUMs (low speed and high torque) and DC (high speed and low torque) motors up to 7–8 W.

Dynamic performance

This section deals with the time response of the actuators and their energetic characteristics. Where direct application of the piezoelectric effect is concerned, the coupling factor gives a good indication of the transduction efficiency. The coupling factor is defined as the square root of the energy ratio between stored mechanical energy and input electrical energy.

For piezoelectric ceramics, again depending on their piezoelectric parameters, the coupling factor is of the order of 0.7–0.75. As in the previous case, particular actuator configurations determine very different dynamic properties.

- *Energy Density and Specific Energy Density.* These are defined as the ratio of output mechanical energy per cycle to actuator size or weight, respectively. From theoretical considerations of maximum breakdown electric field sustained by dielectric ceramics, values of up to about 50 Jcm^{-3} have been reported for the maximum energy density of piezoelectric actuators (Madou (1997)).

  In the case of specific actuator configurations, the theoretical figures are one or two orders of magnitude lower. Depending on the type of actuator, the energy density falls within the range 10^{-3}–10^{-1} Jcm^{-3} (see Figure 2.32).

- *Power Density and Specific Power Density.* Owing to the high dynamic range of piezoelectric actuators, the power density can also be high; however, this depends very much on the actuator configuration. For stack piezoelectric actuators (which most resemble direct application of the piezoelectric effect),

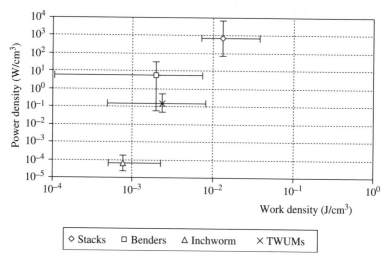

Figure 2.32   Relative position of piezoelectric stacks, benders, inchworm actuators and TWUMs on the energy–power density plane.

the power density is in the range 10^3–10^4 Wcm^{-3}. However, for inchworm actuators, it can be as low as 10^{-1} Wcm^{-3}. Other piezoelectric actuators lie in between these upper and lower limits; for instance, benders perform with a power density of the order of 10 Wcm^{-3}, while TWUMs perform in the range 10^{-1}–1 Wcm^{-3} (see Figure 2.32).

- *Time Constant and Frequency Bandwidth.* As indicated in chapter 1, both time constant and frequency bandwidth are inversely related and to a large extent define the dynamic properties of an actuator. Piezoelectric stacks present time constants ranging from microseconds to milliseconds and producing maximum frequency bandwidths up to hundreds of kilohertz. The frequency range of piezoelectric benders is typically limited to a few kilohertz, and inchworm actuators and TWUMs cannot perform at more than a few tens of hertz.

- *Energetic Efficiency.* Piezoelectric actuators are generally very efficient in converting from electrical to mechanical energy. Piezoelectric stacks, benders and inchworm actuators (whose operational characteristics bring them very close to the concept of direct application of the piezoelectric effect) can attain 70–90% energetic efficiency.

In the case of piezoelectric actuators that rely on frictional transmission of forces, efficiency drops to 30–40%. This is true, for instance, of TWUMs.

The behavior of this technology upon scaling is analyzed in detail in the next section.

2.6.2 Scaling of piezoelectric actuators

In view of current trends toward miniaturization, it is worth inquiring how the performance of piezoelectric actuators is affected by reducing their size. We are not concerned here with the domain of microactuators, that is, actuators with sizes in the micrometer range.

In addition to the intrinsic change of driving characteristics directly related to the actuator, the influence of changes in physical phenomena may be relevant in the domain of microactuators. This is true of surface forces that become dominant as compared to volume forces when the application is scaled to this domain. For a detailed discussion on scaling laws, the reader is referred to excellent works by Madou (1997) or Peirs (2001).

It has been reported that the piezoelectric effect scales down with the size of the actuators but is expected to have a measurable impact on a microscopic scale. The analysis in this section focuses on four useful parameters for describing the performance of actuators:

1. *Resonance frequency.* Resonance frequency is a very important parameter in describing the performance of piezoelectric drives, irrespective of whether they are resonant or nonresonant drives. In resonant drives, it is the resonance frequency that is used to drive the actuators; this is closely related to the speed of the linear or rotative motion and defines the characteristics of the electronic driver to a great extent.

   In nonresonant drives, on the other hand, the resonance frequency is usually the upper limit for the feasible driving frequency.

2. *Stroke.* The stroke is an important parameter in the case of nonfrictional transmission of displacement, for example, in piezoelectric stacks and multimorph benders. In the case of frictional transmission of displacement, for example, linear or rotational ultrasonic motors and inchworm motors, the stroke is either unlimited or it is only limited by the rotor length.

3. *Force density.* The force density describes the ratio of available force to volume or weight of the actuator. It is useful because it is closely related to the time response of the actuator.

4. *Power density.* The power density can be obtained from the previous parameters. It is defined as the ratio of available power to volume or weight.

Resonance frequency

Manufacturers of piezoelectric drives usually give the following relationship between resonance frequency of the actuator and the size:

$$f_r = \frac{N}{L} \tag{2.41}$$

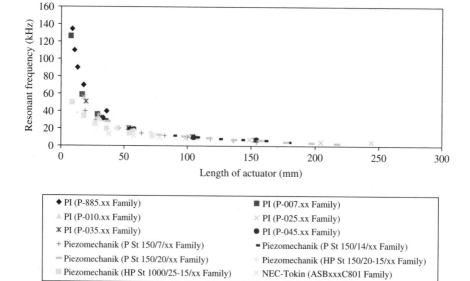

Figure 2.33   Evolution of resonance frequency upon miniaturization of piezoelectric stack actuators.

where f_r is the resonance frequency of the piezoelectric ceramic, N is an actuator-specific constant dependent on the vibration mode of the particular ceramic (e.g. longitudinal for stacks, flexure for multimorphs) and L is the actuator length in the direction of the vibration mode (e.g. length for stacks, thickness for multimorphs).

According to Equation 2.41, all types of piezoelectric drives should exhibit the same tendency for the resonance frequency to increase at a rate inversely proportional to the decrease in size. Figure 2.33 shows the evolution of the resonance frequency of piezoelectric stack actuators from various different manufacturers.

As the figure shows, the overall trend in piezoelectric stacks conforms to Equation 2.41. This result is also consistent with the scaling analysis of Peirs (2001). According to this analysis, the stiffness of second-order mechanical systems scales down linearly with the size of the actuator, that is, $K \propto L$. Since the mass of the actuator will scale down according to the volume of the actuator (i.e. $M \propto L^3$), the resonance frequency of the actuator (which is a second-order mechanical system) is

$$f_r \propto \sqrt{\frac{K}{M}} \propto L^{-1} \qquad (2.42)$$

The result of Figure 2.33 confirms the resonance frequency trend described by Equations 2.41 and 2.42; it also indicates that additional bandwidth is left available for nonresonant drives and that the driving frequency for resonant drives will increase upon miniaturization.

Stroke

As explained earlier, stroke analysis is important for nonfriction-driven piezoelectric actuators. As in the previous case, the trend of the stroke of piezoelectric actuators is independent of the particular type of nonfrictional actuator. The stroke is always linearly related to the dominant dimension in the direction of actuation, for example, length in piezoelectric and in multimorphs and diameter in radial expanders.

Force density

Force density is defined here as the ratio of available force to volume. Force density is closely related to the acceleration that the drive is able to impart to the load and also to the response time of the system.

Since the mass, M, and the volume, V, of the actuators are proportional, the force density is also proportional to the acceleration, a, of the load:

$$M \propto V \quad \rightarrow \quad \frac{F}{V} \propto a \quad (2.43)$$

The experimental relationship between force density and volume is depicted graphically in Figure 2.34. It will be seen that the force density in piezoelectric actuators is inversely proportional to the length in the direction of stroke. This again confirms the theoretical result of Peirs (2001), who established the following relationship:

$$\frac{F}{M} \propto \frac{F}{V} \propto \frac{1}{L} \quad (2.44)$$

where L is the dominant dimension in the actuation displacement.

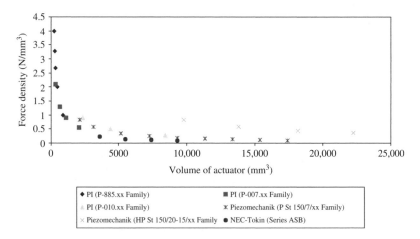

Figure 2.34 Evolution of the force density, $\frac{F}{V}$, of piezoelectric stack actuators upon miniaturization.

Table 2.3   Operational characteristics and scaling trends for piezoelectric actuators

Figures of merit	Nonresonant actuators			Resonant actuators
	Stack	Benders	Inchworm	TWUM
Force, F	10^2–10^5 N	10^{-2}–10^2 N		Torque $\leq 10^4$ Nm
Displacement, S	0.1–0.6% (dynamic up to 1.7%)	10^{-2}–10 mm	Unlimited	Unlimited
Work density, W_V	$\approx 10^{-2}$ J/cm^3	10^{-4}–10^{-2} J/cm^3	$\approx 10^{-3}$ J/cm^3	10^{-3}–10^{-2} J/cm^3
Power density, P_V	10^2–10^4 W/cm^3	10^{-1}–10^1 W/cm^3	$\approx 10^{-4}$ W/cm^3	$\approx 10^{-1}$ W/cm^3
Bandwidth, f	$\leq 10^5$ Hz	$\leq 10^3$ Hz	$\leq 10^2$ Hz	$\leq 10^2$ Hz
Efficiency, η		70–90%		
Scaling trends				
Force	$F \propto L^2$			
Stroke	$S \propto L$			
Work per cycle	$W \propto L^3$			
Energy density	$W_V \propto L^0$			
Bandwidth	$f \propto L^{-1}$			
Power density	$P_V \propto L^{-1}$			

A similar analysis can be used to establish the trends in response time of the piezoelectric actuator upon miniaturization. It is clear from Equations 2.43 and 2.44 that the response time tends to decrease linearly upon miniaturization:

$$a \propto \frac{1}{L} \propto \frac{L}{T^2} \quad \rightarrow \quad T \approx L \qquad (2.45)$$

This analysis takes only the mechanical characteristics of the active material into account. In the derivation, the volume of the electronic drive was not taken into consideration so that they indicate trends rather than the exact situation.

This can be seen again in the case of the response time. As explained earlier, one of the factors limiting the response time derives from the charging and discharging time of the capacitor that piezoelectric actuators represent when driven out of resonance. The electrical capacitance of the piezoelectric actuator, C_p, is proportional to the capacitor area, A, and inversely proportional to the distance between electrodes, L. The tendency of the electrical capacitance would be to decrease linearly when the actuator is miniaturized:

$$C_p \propto \frac{A}{L} \propto L \qquad (2.46)$$

This will produce an effect on the response time in addition to the effect discussed in the foregoing paragraphs. See Table 2.3 for a summary of the scaling trends and figures of merit of piezoelectric actuators.

2.7 Applications

2.7.1 Applications of resonant piezoelectric actuators

Resonant piezoelectric motors have reached a mature stage of development. The most successful implementation is the travelling wave ultrasonic motor. There are several ultrasonic motors available off the shelf on the market. The first application considered in this section is that of OEM ultrasonic motors from Shinsei Corporation Inc., Japan. This is followed by a brief description of a second application, which is the implementation of ultrasonic drives in the Canon optical lens automatic focus.

Case Study 2.1: OEM Ultrasonic motors, USR-60 and USR-30
(Shinsei Corporation Inc., Japan)

The principle of operation of ultrasonic motors was described in detail in Section 2.3.2. The first commercially available ultrasonic motor appeared as early as 1986, with the product name USR-4-100. This was the precursor of USR-60, the current commercial name of Shinsei's 60-mm-diameter ultrasonic motor.

Practical applications for ultrasonic motors have been developed in a wide range of fields: auto-focus optical lenses in cameras, lens-mirror actuation in optical devices, positioning in satellite reception devices, winding-up function in roll screens, and headrest adjustment in automobiles.

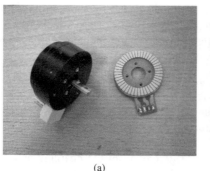 

(a)   (b)

Figure 2.35   Detailed view of (a) a USR-30 and (b) a USR-60 motor from Shinsei.

As explained earlier, ultrasonic motors present low-speed operation at high torque. This can be beneficial in applications where compact solutions are required (e.g. avoiding transmission stages in a direct drive approach).

The ultrasonic driving frequency, always above 20 kHz, ensures silent operation. In addition, the static torque serves as a practical self-braking mechanism owing to the frictional transmission.

The USR-30 and USR-60 (see Figure 2.35) share the same motor structure. They are two-phase drives. The piezoelectric ceramic exciting the microscopic oscillation in the stator is poled in an alternating pattern providing a *Sine* and a *Cosine* mode. See Section 2.3.2 for more details.

An active sector of the piezoelectric ceramic is used as a feedback sensor for the electronic driver. The piezoelectric ceramic sector serving as a sensor is poled similarly to the driving sectors. The electrode in the sensor sector picks up an electrical signal, which is proportional to the vibration amplitude in the stator.

Ultrasonic motors are controlled by appropriate modification of the driving frequency, and, therefore, speed variations are directly related to modifications of the vibration amplitude at the stator. The sensor signal is thus compared to the reference velocity and to an error signal, which serves as a control for the driving frequency. See the schematic representation of the driver in Figure 2.36.

This approach to driving is equivalent to the tracking process for resonant drives described in previous sections. The main difference lies in the use of an additional embedded sensor rather than a phase lag between the driving voltage and the current drawn to track the resonance frequency.

Both the USR-30 and the USR-60 come in a number of optional configurations. In particular, twin-shaft and single-shaft options are available together with versions with optical encoders.

Case Study 2.2:   Ultrasonic motor focus in EF-28-105 USM Lens

Ultrasonic motors provide an elegant and amazingly compact solution to the automated optical focus in reflex cameras. Canon was the first company to introduce

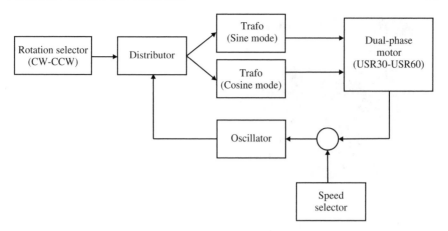

Figure 2.36   Control block diagram of USR-60 driver. The purpose of the feedback signal from the embedded sensors is to keep the output rotation speed close
to the reference.

ultrasonic motors in their reflex cameras. Initially, the 77-mm L–1 ultrasonic motor
was used to drive the electronic focus of 50 f/1.0 and 1200 f/5.6 lenses.

The next version was the well-known 62 mm M–1 USM (Ultrasonic motor),
which became the driving technology for the automatic focus in 28–105 f/3.5–4.5
and 85 f/1.8, 300 f/4L and 70–200 f/2.8L lenses.

Yet a third, miniaturized version of the two previous ultrasonic motors, the
Micro-USM, is used to drive the 28–80 II, III and IV and the 50 f/1.4.

Optical lenses benefit from the simple structure of the TWUM principle. The
above three implementations consist in a hollow toothed stator to prevent obstruc-
tion of axial optics, and a ring rotor. Both are pressed together by a spring-loaded
bayonet-mounted plate.

The operating principle is well-known: microscopic travelling vibrations excited
at the stator's teeth are transmitted to the rotor through frictional mechanisms. The
low time constant of ultrasonic motors ensures fast, smooth and silent operation.
Figure 2.37a is a view of a Canon optical lens and Figure 2.37b is a schematic,
depicting the configuration of the 62 mm M–1 USM Ultrasonic Motor.

2.7.2   Applications of nonresonant piezoelectric actuators

The traditional field of application of piezoelectric actuators is in precision positioning stages. There are very many manufacturers worldwide providing micropositioning and nanopositioning stages for application domains such as astronomy,
semiconductor testing systems, medical engineering, biotechnology and telecom-
munications to mention only a few.

The application domain of machine tool systems is in the full flow of evolution.
At this time, traditional material removal processes are being used to provide

      (a)   (b)

Figure 2.37   Canon lens including a travelling wave ultrasonic motor: (a) view of the lens and (b) schematic view of the 62-mm M–1 USM.

surface finishing and dimensional accuracy in the nanometer range. In order to achieve such high specular finishes, the machine tool must ensure high stiffness and the high accuracy required for precise positioning between tool and workpiece in all directions.

Case Study 2.3:   Stepping piezoelectric motors for high-accuracy, high-stiffness machine tools

Our first example was developed by the Production, Machine Design and Automation Division of the Department of Mechanical Engineering at the Katholieke Universiteit Leuven (PMA-KULeuven). It consists in a nanometer-precision, ultra-stiff piezoelectrically driven stage for ELID (Electrolytic In-process Dressing) grinding. The text and pictures illustrating this example are provided courtesy of PMA-KULeuven.

Existing machine tools provide insufficient stiffness to guarantee high-precision positioning in the presence of varying process forces. To illustrate this problem, consider that where cutting forces perpendicular to the workpiece surface are of the order of 10 N, stiffness must be as high as 1 kN/µm to ensure flatness of the order of 10 nm.

The use of piezoelectric stepping motors to provide a system that is able to combine the guiding and actuation requirements (i.e. the mechatronic approach: combination of several functions on the same component) for machine tools is an innovative solution that guarantees extreme stiffness between workpiece and tool in all directions.

The operational requirements for the drive-guide stage are

1. High stiffness

2. Smooth motion.

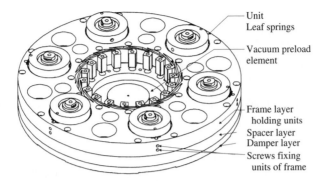

Unit
Leaf springs

Vacuum preload
element

Frame layer
holding units
Spacer layer
Damper layer
Screws fixing
units of frame

Figure 2.38   Ultrastiff positioning stage comprising six positioning units (Courtesy of Dominiek Reynaerts. Reproduced by permission of PMA-KULeuven).

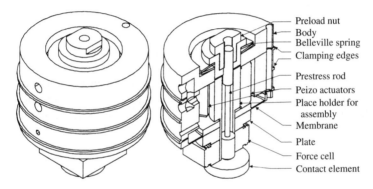

Preload nut
Body
Belleville spring
Clamping edges

Prestress rod
Peizo actuators
Place holder for
assembly
Membrane

Plate

Force cell
Contact element

Figure 2.39   Positioning stepper unit comprising three piezoelectric stack actuators (Courtesy of Dominiek Reynaerts. Reproduced by permission of PMA-KULeuven).

The proposed stage consists of a circular plate supported by six piezoelectric stepping units. The stage shown in Figure 2.38 can be actuated to three degrees of freedom (x, y and C) by moving the six piezoelectric units according to a predefined gait pattern.

The structure of each unit is depicted in Figure 2.39. This consists of three piezoelectric stack actuators, the elements required to ensure preloading of the actuators, force sensors to provide feedback on the contact force and a spherical contact element with a large radius to establish contact between the unit and the ground.

The "hammering" effect characteristic of stepping drives can be overcome in the proposed configuration by simultaneously extending all piezoelectric stacks according to the locomotion principle depicted schematically in Figure 2.40. The piezoelectric stacks are driven in a cyclical pattern but well below their resonance frequency, which places them in the category of nonresonant drives.

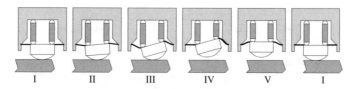

Figure 2.40 Operational cycle of a single positioning unit (Courtesy of Dominiek Reynaerts, Reproduced by permission of PMA-KULeuven).

The hammering effect is due to the characteristic discontinuity of the stepping motion. It is chiefly a function of two main parameters: (i) the distance between each spherical contact surface and the ground; and (ii) the load to be supported by the foot.

In order to overcome the hammering problem, a stepping algorithm was implemented on the stepping controller. This stepping algorithm is based on an initial estimate of the distance between driving unit and ground and of the load to be supported. A stepping function is generated on the basis of this estimate. The stepping function is such that there can be no positional deviation of the stage if the two parameters were estimated correctly.

The controller implements an update of the stepping function on the basis of the detected positional error and may thus be described as a feed-forward learning controller. The hierarchical control scheme is shown in Figure 2.41. The main

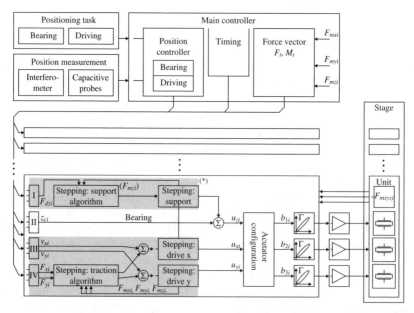

Figure 2.41 Schematic representation of the modular controller (Courtesy of Dominiek Reynaerts. Reproduced by permission of PMA-KULeuven).

controller implements the control signal on the basis of the reference and actual positions as described in the task programming module. As a result, instructions are broadcast to each unit controller.

The unit controller includes its own amplifier and contains a stepping algorithm for the z direction and two traction algorithms for x and y directions respectively. The hysteresis in the piezoactuator is compensated by an inverse model so that the unit presents a linearized input–output behavior.

To illustrate the performance of the ultrastiff precision positioning stage, Figure 2.42 shows the results of tracking a circle with a diameter of 1 mm in the xy-plane. Likewise, Table 2.4 shows some of the stiffness and driving characteristics of the machine tool stage.

The piezoelectrically driven stepping stage has been tested with the forces resulting from an ELID grinding process. The ELID grinding process results in a pattern with an average force between 10 and 30 N having a superimposed time-varying component with lower amplitude and low frequency (around 25 Hz and some harmonics). The positional deviation due to this process force is less than 20 nm when the active stiffness is implemented.

Case Study 2.4: Bending trimorph actuators for needle selection modules in knitting machines

One of the leading manufacturers of piezoelectric bending elements is **Argillon GmbH**, formerly SIEMENS AG, Ceramics GmbH. Their bending elements are

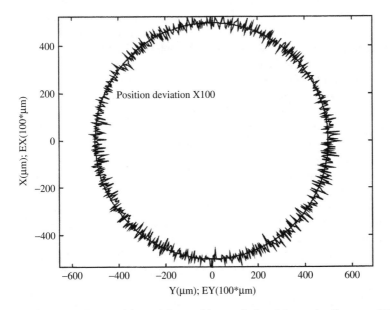

Figure 2.42 X- and Y-position while tracking a circle of 1 mm in diameter (Courtesy of Dominiek Reynaerts. Reproduced by permission of PMA-KULeuven).

Table 2.4   Performance parameters (positioning and
stiffness) of the ultrastiff positioning stage (Courtesy
of Dominiek Reynaerts, Reproduced by permission
of PMA-KULeuven)

Feature	Drive	Bearing
Positioning		
Resolution	5 nm	2.5 nm
Measurement accuracy	0.3 μm	0.1 μm
Stroke	100 mm	5 μm
Velocity	2 mm/s	–
Stiffness		
Passive (3 units)	55 N/μm	150 N/μm
Active (3 units, <50 Hz)	120 N/μm	320 N/μm
Passive (5 units, <50 Hz)	70 N/μm	200 N/μm
Active (5 units, <50 Hz)	150 N/μm	450 N/μm

made to the trimorph concept described above, and they are therefore suitable for
fast actuation in the range of attractive blocking forces and relatively high (as
compared to stacks) displacements.

Argillon has introduced piezoelectric benders in several industrial sectors. In
particular, they are used for Braille equipment, textile machines, pneumatic valves
and medical equipment. The case study presented here uses Argillon trimorph actuators (conceptually similar to the bimorph configuration) for fast needle selection
modules in knitting machines. The text and pictures in this application case are
courtesy of Argillon GmbH.

Traditionally, control of knitting machines is based on punch cards or magnets.
Piezoelectric benders offer a number of advantages as compared to traditional
control modules. In addition to faster and more reliable operation, they provide
high efficiency. Faster and more reliable operation results in increased productivity,
owing to the shorter switching times achieved with piezoactuators.

In addition to improved dynamic performance, the control units are so small
that the space requirements on the machine are minimal. Since the piezoelectric
bender operates on a low energy supply and the overall efficiency is high, no
cooling units are required so that even more compact solutions are possible.

Argillon's trimorph technology is conceptually similar to the bimorph configuration described in previous sections. The trimorph bender implemented in the
needle selection unit measures 49.95 mm in overall length, 7.20 mm in width and
0.80 mm in total thickness.

The thickness of the intermediate carbon fiber vane (the carrier) is 0.24 mm.
The overall maximum deflection of each piezoelectric bender is 1.70 mm at 200 V
(38 mm free length), while the blocked force is just 0.7 N. The trimorph actuator

Figure 2.43   Needle modules for knitting machines (Courtesy of U. Zipfel. Reproduced by permission of Argillon GmbH).

is set up in a parallel configuration direction for both piezoelectric ceramics is opposite and the internal electrode is connected to a reference voltage (ground).

Figure 2.43 shows the trimorph arrays in three different needle selection modules. Each actuator exhibits a typical capacitive load of 43 nF and can be operated at a maximum voltage of 230 V.

Case Study 2.5:   Precision piezoelectric XYZ scanning mechanism for atomic force microscopy on a spacecraft

The MIDAS (Micro Imaging Dust Analysis System) instrument was launched on March 2004 on board the European Space Agency probe ROSETTA (Ariane 5 mission). The ROSETTA/MIDAS system will analyze the dust of the 46/P Churyumov – Gerasimenko comet using an atomic force microscope.

Cedrat Technologies (Meylan, France) has designed and implemented a piezoelectric XYZ stage that will be used to scan the dust specimen on three axes. The caption and the pictures of this instance of application are courtesy of Cedrat Technologies. The scanning system, a constituent part of the complete instrument, has been qualified for EQM (engineering and qualifications model), QM (qualification model), FM (flight model) and FSM (flight spare model) standards.

The scanning mechanism contains six degrees of freedom. Three of these are actively controlled (positions X, Y and Z), and three passively cancelled (rotations around axes X, Y and Z). The target functional performance was specified as follows:

- Scanning stroke in X, Y directions: 100 µm,

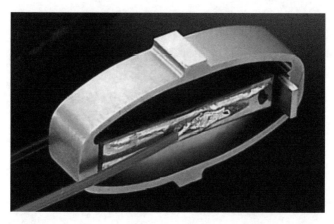

Figure 2.44   CEDRAT TECHNOLOGIES' APA50S piezoelectric actuator (Courtesy of R. Le Letty and F. Claeyssen. Reproduced by permission of CEDRAT TECHNOLOGIES).

- Scanning stroke in Z direction: 8 μm,

- Maximum parasitic rotation: $\theta_z \leq 240$ μrad,

- Maximum parasitic rotation: $\theta_x, \theta_y \leq 20$ μrad.

The fundamental part of the scanning stage is Cedrat's standard Amplified Piezoelectric Actuator (APA50S) (see Figure 2.44 for a detailed view of the actuator). The configuration of the XY stage is symmetrical and it uses eight actuators to achieve a travel of 100 μm.

As well as meeting the application requirements in terms of stroke and robustness, amplified piezoelectric actuators APAs were chosen because they are simpler and easier to build than competing technologies. This is important for the possibility of integrating position sensors for each scanning drive direction. The displacement in each direction is monitored by means of a capacitive sensor.

Moreover, the APAs are integrated in a parallelogram configuration and so act additionally as guiding elements. Flexural hinges are implemented to decouple the displacement in X and Y directions (see Figure 2.45).

The scanning motion in the Z direction is provided by a customized parallel prestressed actuator with a stroke of 8 μm. The prestressed actuator is equipped with a full Wheatstone bridge of strain gauges to monitor the Z displacement.

A common aspect of spacecraft applications is the strictness of requirements in terms of shock and vibration during the launching stage. In order to solve this problem, the scanning stage was equipped with a latching mechanism based on shape memory alloys (see Case study 3.1 where the latching mechanism is described).

During the testing phase at CEDRAT TECHNOLOGIES' laboratory, the XY stage was excited using CEDRAT TECHNOLOGIES' SA75 electronic drivers,

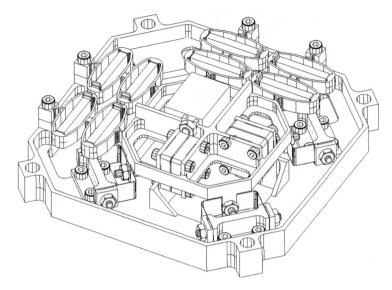

Figure 2.45 I-DEAS view of the XY stage (EM) (Courtesy of R. Le Letty and F. Claeyssen. Reproduced by permission of CEDRAT TECHNOLOGIES).

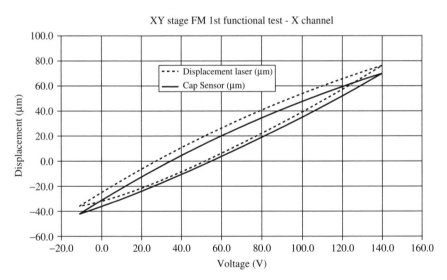

Figure 2.46 Low hysteresis behavior of the XY stage (Courtesy of R. Le Letty and F. Claeyssen, Reproduced by permission of CEDRAT TECHNOLOGIES).

Figure 2.47   View of a piezoelectric XYZ stage flight model for the Rosetta/Midas instrument (Courtesy of R. Le Letty and F. Claeyssen, Reproduced by permission of CEDRAT TECHNOLOGIES).

and the response from the capacitive sensors was compared to that of a laser interferometer. The linearity error in the displacement versus voltage characteristic of the scanning stage was less than 0.5% (see Figure 2.46). The ratio between the no-load displacement of APAs and the stage displacements was about 0.88 and was related to the increased stiffness of the elastic flexure hinge–based guiding mechanisms. Also, the coupling between X and Y axes was less than 2%.

Figure 2.47 shows the flight model of the scanning mechanism for the ROSETTA–MIDAS instrument. The model includes the XYZ positioning stage, the capacitive and strain gauge position sensors and the latching mechanism based on shape memory actuators.

3

Shape Memory Actuators (SMAs)

Shape Memory Actuators, SMAs, are devices that make use of the *shape memory transformation*. The first reported shape memory transformation was by Chang and Read, who observed its reversibility in 1931. It was not until 1951 that the shape memory effect was demonstrated experimentally in a A_uC_d bent bar.

The development and application of shape memory alloys attained the requisite momentum following the discovery of N_iT_i alloys in 1963 at the Naval Ordnance Laboratory. The name *nitinol* refers to **Ni**ckel, **Ti**tanium, **NOL** (Naval Ordnance Laboratory). The implementation of applications making use of SMAs has evolved hand-in-hand with the development of *nitinol*.

In the last few decades, both industry and academia have evinced growing interest in the application of nitinol, and of SMAs in general. This is basically because these materials are intrinsically susceptible of use both as sensors and as actuators, which makes them suitable for use as smart actuators and for integration in smart structures.

As in many other cases, side-by-side development of material fundamentals and applications is of paramount importance for the successful development and implementation of SMAs. So far, the application of shape memory alloys as actuators is only a small proportion of the nitinol in use. Most of the current applications of nitinol make use of the so-called superelastic effect in shape memory alloys.

The applications in the medical field rely on the so-called superelastic or pseudoelastic behavior of shape memory alloys. They also take advantage of the good corrosion and biocompatibility characteristics of nitinol. Solutions are comparatively simpler in a passive approach using the superelastic characteristics of nitinol than in an active approach using the shape memory effect.

Emerging Actuator Technologies: A Micromechatronic Approach   J. L. Pons
© 2005 John Wiley & Sons, Ltd

This chapter provides a detailed description of the shape memory effect and the superelastic effect in shape memory alloys. The discussion focuses initially on the characteristics of the shape memory transformation. This discussion then provides the basis for a detailed description of the design and use of shape memory actuators, in which material, thermal, electrical and mechanical considerations are addressed.

A specific section of the chapter is devoted to the control of shape memory actuators as approached from a mechatronic point of view. This includes a description of the sensing capability of shape memory actuators and discussion of the use of SMAs as smart materials.

The closing part of the chapter discusses scaling effects in shape memory actuators and their actuation characteristics. The use of SMAs is exemplified by descriptions of several representative case studies.

3.1   Shape memory alloys

Shape memory alloys exhibit a thermally activated *martensitic transformation*. In shape memory alloys, there are two stable phases: *martensite* (also referred to as α-phase), and the *parent phase* or *austenite* (β-phase).

The parent phase in SMAs is only stable above a given temperature A_s, known as the reverse transformation start temperature, and is the only stable phase once the so-called reverse transformation finish temperature, A_f, is reached. Similarly, the martensite phase is only stable below a given temperature, M_s, the martensite start temperature, and again is the only stable phase at temperatures below the martensite finish temperature, M_f (see Figure 3.1).

The martensitic transformation is a *reversible, diffusion-less* transformation. It is reversible in that upon heating from the low-temperature martensite phase, the onset of the parent phase will commence above A_s and will be complete

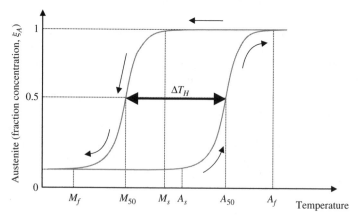

Figure 3.1   Transformation temperatures in shape memory alloys.

when the temperature exceeds A_f. Upon cooling to below M_s from the high-temperature parent phase, the transformation back to martensite will commence. That transformation will be complete when M_f is reached.

The martensitic transformation is diffusion-less: that is, neighboring atoms in the lattice will remain in equal relative positions after the thermally driven transformation, even though the relative distance will be altered. In addition, at any given temperature, the fraction concentration of each phase will remain constant and independent of time.

A fundamental difference between the martensite and parent phases, and the basis for an explanation of the *shape memory effect*, is crystallographic symmetry. The parent phase exhibits greater symmetry than the martensite phase. Upon cooling from the parent phase, the lack of symmetry may lead to the formation of up to 24 martensite *variants* or *domains*. In fact, it is common for two to four variants of martensite to form side by side with the function of relaxing the elastic strain energy around the variants. The process whereby this relaxation is achieved is known as *self-accommodation* of variants (see Figure 3.2).

Along with the lower symmetry in the martensite phase, the other factor responsible for the so-called *stress-induced accommodation* of martensite variants is the high mobility of the interface between these variants. The stress-induced accommodation of martensite variants is responsible for the low stress required to achieve a macroscopic deformation of the material when in the martensite phase (see Figure 3.2).

3.1.1 The shape memory effect

The shape memory effect is a characteristic phenomenon of thermoelastic transformations. It is the outcome of three combined effects:

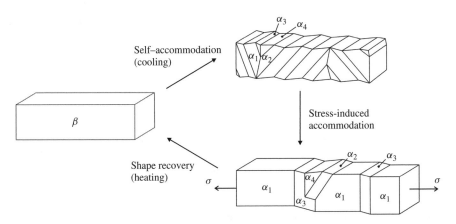

Figure 3.2 Illustration of the self-accommodation process upon cooling from the parent phase, the stress-induced accommodation of variants in the martensite phase and shape recovery after heating from the martensite phase.

1. Self-accommodation of martensite variants to minimize strain energy upon transformation from the parent phase.

2. Stress-induced accommodation of twin related martensite variants.

3. Difference in symmetry between parent and martensite phases.

The shape memory effect is schematically illustrated in Figure 3.2. The ordered symmetrical crystallographic structure shown in grey represents the high-temperature $(T \geq A_f)$ parent phase. Upon cooling below M_f, the alloy is transformed into martensite domains that undergo what is called a self-accommodation process to reduce strain energy. The result is a low-symmetry structure with twin related martensite variants (shown in black in Figure 3.2), with no significant macroscopic shape change as compared to the parent phase.

When a mechanical stress is applied to the alloy in the martensite phase, there is stress-induced accommodation of variants, causing a reorientation of twin related variants and resulting in a macroscopic deformation (see the schematic in Figure 3.2). When the material is heated again, the martensite variants that were accommodated when stress was applied will revert to the original orientation in the parent phase and the original shape will be recovered.

The thermomechanical behavior of an alloy exhibiting the shape memory effect is depicted schematically in Figure 3.3. Figure 3.3 shows the three-dimensional

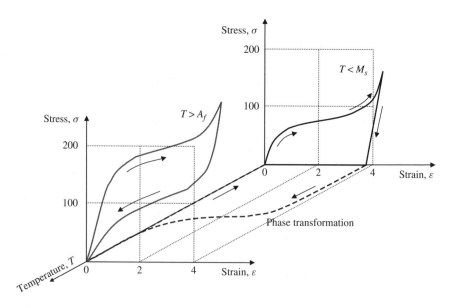

Figure 3.3 Thermomechanical behavior of shape memory alloys. Low-stiffness behavior in the martensite phase due to stress-induced accommodation of martensite variants (black line), shape memory effect during phase transformation (dashed line) and superelastic effect at temperatures above A_f (grey line).

relationship between stress, strain and temperature for a shape memory alloy. While the material is at low temperature, $T \leq M_f$, (see black curve in Figure 3.3), the application of a stress will cause stress-induced accommodation of twin related α variants, resulting in a permanent macroscopic deformation of the order of a few percentiles, up to 6–8%.

At this point, when the material is heated up from the martensite phase (see black line in Figure 3.3), the transformation to the parent phase will commence. Transformation will be complete $T \geq A_f$. Owing to the greater symmetry of this phase, the alloy will revert to the original shape (the one it had prior to the stress-induced deformation in the martensite phase). This process of shape recovery upon heating to the parent phase is the basis of shape memory actuation. It can be appreciated better in the 2D projection in Figure 3.4.

Once in the high-temperature parent phase, the alloy can be cooled down to the martensite phase without any shape change. For the alloy to be used in another actuation cycle, stress-induced deformation would be required in the martensite phase. This is the basis of what is known as the *one-way shape memory effect, OWE*.

Transformation hysteresis and range

$N_i T_i$ alloys do not complete their phase transformation at one particular temperature. The martensitic transformation in either direction begins at one temperature (known as the start temperature) and is completed at another temperature (known as the finish temperature) (see Figure 3.1).

There is a difference in the transformation temperatures when the temperature is increased from martensite to austenite and when the material is transformed from

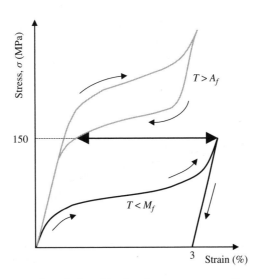

Figure 3.4   Stress–strain relationship in shape memory alloys and process of shape recovery.

austenite to martensite, resulting in hysteresis in the transformation. This difference is known as the *transformation temperature hysteresis* and is shown in Figure 3.1 as ΔT_H.

The transformation hysteresis can be defined as the difference between the temperatures at which the material is 50% transformed to austenite upon heating and 50% transformed to martensite upon cooling. Typical values for binary NiTi alloys are about 25 to 35 °C.

The hysteretic behavior of shape memory alloys can be explained (i) in terms of the free enthalpy (G) associated with each stable phase (martensite and austenite); (ii) in terms of the equilibrium temperature for which the free enthalpy for both phases is matched and (iii) occasionally in terms of the overheating required to promote the transformation from martensite to austenite and the undercooling required to promote the reverse transformation. This is illustrated qualitatively in Figure 3.5.

The fraction concentration of phases in the alloy is such that the free enthalpy is minimized. There is an equilibrium temperature, T_{eq}, at which the free enthalpy of both the parent and the martensite phases are equal. However, the transformation from the parent to the martensite phase requires cooling of the material to below the equilibrium temperature. Similarly, the transformation to the parent phase requires overheating above the equilibrium temperature. This situation, illustrated in Figure 3.5, causes hysteretic behavior of the strain versus temperature relationship in shape memory alloys.

The width of the hysteresis in the strain versus temperature relationship is a function of several factors, namely, the relative content of alloying elements (Ni and Ti in the case of nitinol), the presence of ternary elements in the alloy composition, the alloy microstructure resulting from thermomechanical treatments and, occasionally, the application. The effect of ternary additions to the base alloy will be discussed in more detail in the following sections.

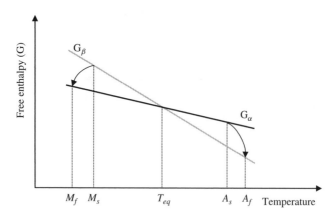

Figure 3.5   Free enthalpy as a function of transformation temperatures in shape memory alloys.

It is worth noting that, in addition to the effect on the strain–temperature relationship, all the above-mentioned parameters invariably affect the overall electromechanical properties of the alloy and the transformation.

The absolute position of the transformation temperatures and the overall span of the transformation can also be important in addition to the hysteresis. This is true, for instance, of application in the automobile industry. The automotive application requires higher transformation temperatures ($M_f \geq 85\,^\circ$C), which are not currently available in commercial alloys.

Moreover, if the application requires complete transformation upon both heating and cooling, then the difference between A_f and M_f must be considered. Typical values for the overall transformation temperature span are about 40 to 60 °C.

Changes in electrical, mechanical and chemical characteristics concomitant to the shape memory effect

Shape recovery and pseudoelasticity are probably the best-known macroscopic changes accompanying the phase transformation in shape memory alloys. Nevertheless, a significant number of other physical and chemical properties are also altered during the transformation.

Harrison (1990) reported an extensive analysis of the different properties subject to modification during the martensite transformation:

- *Electrical Properties*. Electrical resistivity, the thermoelectric power, electromotive force, magnetic and other properties are modified during the transformation process. In particular, electrical resistivity enables shape memory alloys to be used as sensors.

- *Mechanical Properties*. Yield strength, Young's modulus, damping and internal friction are some of the mechanical properties that are altered during the transformation. The change in the Young's modulus is of particular interest where smart actuators and smart structure applications are concerned.

- *Thermal Properties*. During the transformation, the thermal conductivity, the heat capacity and the latent heat of transformation are also altered.

Because of the different changes undergone by the material during transformation, shape memory alloys are one of the best suited materials for developing the concept of concomitant sensing and actuation: that is, smart actuators (see Section 3.3.2). Also, the changes in mechanical and electrical characteristics of SMAs favor their integration as functional elements, both sensors and actuators (see Section 5.4.2).

The two-way shape memory effect, TWE

The effect commonly observed in shape memory alloys – characterized by the recovery of a preset shape upon heating above A_f and the return to an alternate shape upon cooling down M_f – is known as *two-way shape memory effect,*

TWE. Two-way memory is unique in that the material "remembers" different high-temperature and low-temperature shapes.

Creating two-way memory in NiTi alloys involves a complex training process in which one or a combination of the following approaches is used (see Duering *et al.* (1990) for more details):

1. Overdeformation while in the martensite phase

2. Shape memory cycling (cool → deform → heat → repeat)

3. Pseudoelastic cycling (load → unload → repeat)

4. Constrained temperature cycling of deformed martensite.

The TWE after the training process is illustrated schematically in Figure 3.6. The training process induces a dominant orientation of variants upon cooling, and this produces the effective difference in shape between α and β phases.

However, the amount of recoverable strain is generally about 2%, which is much lower than is achievable in one-way memory (6 to 8%). Long-term fatigue and stability performance is not well-known. The inherent temperature hysteresis between the heating and cooling transformations remains; moreover, the transformation forces are extremely low after cooling, and the memory can be erased with only very slight overheating (as low as 250 °C).

3.1.2   Pseudoelasticity in SMAs

In the previous section, the martensitic transformation was described in terms of the four transformation temperatures. However, the transformation temperatures

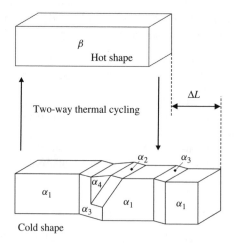

Figure 3.6   Two-way shape memory effect: the material is trained to adopt a dominant orientation of α variants.

for a given alloy are not constant, as they tend to shift toward higher values when a mechanical stress is applied.

The effect of shifting of the transformation temperatures for a given alloy when stress is applied is fully defined by the gradient of the transformation temperature-versus-stress relationship. This gradient is formulated mathematically according to the Claussius–Clapeyron expression, 3.1:

$$\frac{\mathrm{d}\sigma}{\mathrm{d}T} = \frac{\rho \Delta H}{T_0 \Delta \epsilon} = \frac{1}{C_m} \tag{3.1}$$

where ρ is the alloy density, ΔH is the transformation latent heat, T_0 is the transformation temperature (M_s, M_f, A_s and A_f) with no applied stress and $\Delta \epsilon$ is the strain due to the phase transformation. C_m is usually known as the stress gradient. For a graphical representation of this stress-induced transformation temperature shift, see Figure 3.7.

A look at Figure 3.7 shows that the application of stress at a constant high temperature while the material is in the parent phase would lead to the shifting of transformation temperatures and, in some cases, to partial or even total formation of martensite. This phenomenon is usually called *stress-induced martensitic transformation* and produces what is known as *superelasticity*.

The isothermal mechanical effect of the superelastic phenomenon in shape memory alloys is illustrated in Figure 3.4. When the material is in the parent phase ($T \geq A_f$), the application of a mechanical stress will first cause an elastic behavior (see the grey curve in Figure 3.4). Further loading will bring the onset of stress-induced martensitic transformation. Twin related variants in the stress-induced martensite will be subject to stress-induced accommodation, producing a plateau in the stress–strain relationship.

Since martensite is not stable at high temperatures, upon unloading, the parent phase will be fully recovered along with the macroscopic deformation (which may be as much as 10%).

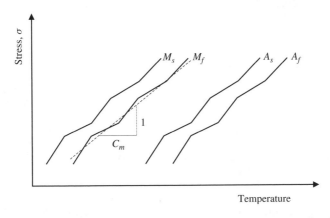

Figure 3.7   Stress-induced shift in transformation temperatures for SMAs.

3.2   Design of shape memory actuators

According to the classification of Duering *et al.* (1990), shape memory alloys have been used in the following types of applications:

1. *Free recovery*. This category encompasses all applications in which the goal is to achieve a net displacement with no significant countervailing external forces; therefore, $\delta F = 0$, and there is no work output.

2. *Constrained recovery*. In this case, the objective is to apply a net force without significant displacement. The use of shape memory alloys for tight coupling of pipes (see Figure 3.8) would come under this heading. Again, $\delta x = 0$, and no net work is produced.

3. *Work-producing devices*. In the two previous cases, there is no significant work output, owing to the lack of either significant displacement or significant force. This category encompasses all applications in which there is a net output of work, $\delta W \neq 0$.

4. *Storage of mechanical energy*. The three previous categories make use of the shape memory effect. The applications in this category, on the other hand, make use of the superelasticity of shape memory alloys, so that superelastic energy is repeatedly stored and delivered, or the energy is dissipated because of the intrinsic hysteretic behavior of superelasticity.

Our analysis of shape memory actuators will focus on applications in the third category. In this category, the shape memory actuator is required to develop a work output by producing displacement against an external load. There are a number of issues that have to be addressed when considering applications of this type. The following sections deal with aspects relating to the amount of work to be delivered, required displacement, bias loading for stress-induced deformation, lifetime, transformation temperatures and span, and heating and cooling processes.

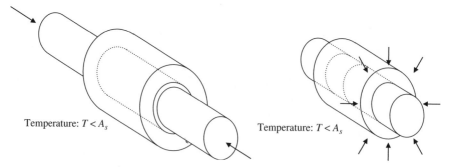

Temperature: $T < A_s$

Temperature: $T < A_s$

Figure 3.8   Tight fitting in permanent couplings exploiting the shape memory effect.

3.2.1 Design concepts for actuation with SMAs

This section outlines various different concepts for designing shape memory actuators, particularly the use of both one-way and two-way memory effects. This leads on to the question of how different types of mechanical loads can be utilized to develop shape memory actuators. This further development of the concept of one-way effect actuators includes a review of aspects relating to available force and displacement, transformation temperatures and fatigue limits.

OWE versus TWE actuators

As noted earlier, the "training process" responsible for the two-way effect in shape memory alloys enables the actuator to be used without the need of any external force to restore the initial shape in the α-phase. This is achieved at the cost of lower recoverable deformations (about 2%) and very limited (virtually zero) force availability upon heating.

In view of these restrictions, it is better to modify the device design, making use of one-way memory by means of a restoring force that acts against the shape memory element to return it upon cooling. Such an approach addresses all of the above limitations. However, the inherent temperature hysteresis remains in both TWE and OWE solutions. Two-way actuators using one-way shape memory elements to act against bias forces generate large strains and high forces in both heating and cooling directions and have demonstrated excellent long-term stability up to hundreds of thousands of cycles.

Mechanical load concepts

Much has been discussed in the previous paragraphs about the ability of shape memory alloys to "remember" the initial shape they had prior to deformation induced by heating to the parent phase. But nothing has been said so far about types of recoverable deformation.

In principle, shape memory actuators can be used under tensile, bending and torsion deformations or under a combination of all three. However, the performance of shape memory actuators depends to a large extent on the particular load case involved. To illustrate this dependence, let us analyze the implications of using each load case.

In general, a material that is subject to normal stress will tend to undergo volume change. On the other hand, where the material is subject to shear stress, the volume of the specimen will remain unchanged. The material will resist normal stresses by means of the material bulk modulus, K, which is defined by the following expression:

$$K = \frac{Y}{3(1 - 2v)} \tag{3.2}$$

where Y is the Young's modulus of the material and v is the Poisson ratio. The Poisson ratio for nitinol, the most commonly used shape memory alloy, is very close to $1/3$, and, thus, $K \approx Y$.

On the other hand, shear stresses are resisted by the material's shear modulus, which is defined in terms of the Young's modulus and Poisson ratio by the following expression:

$$G = \frac{Y}{2(1+v)} \tag{3.3}$$

Assuming the typical value for the Poisson ratio for nitinol, the shear modulus will be $G \approx 3/8 \cdot Y$. In the following paragraphs, we will analyze the three possible load cases most commonly encountered in the design of shape memory actuators.

- *SMAs Subject to Tensile Deformation.* Let us consider the schematic representation of a thin shape memory alloy wire of circular cross section as depicted in Figure 3.9a. When the wire is subjected to a pure tensile stress, the stress distribution over the cross section is constant, σ, and equal to the ratio of the tensile load, P, on the cross-sectional area, A:

$$\sigma = \frac{F}{A} \tag{3.4}$$

For this particular load case, all the material is subject to the same stress level, and, thus, the ratio of the average stress across the cross section, $\hat{\sigma}$, to the maximum stress, σ_{max}, is one to one:

$$\frac{\hat{\sigma}}{\sigma_{max}} = 1 \tag{3.5}$$

Since the maximum strain in the material will, in practice, be the limit to the maximum acceptable tensile load, P_{max}, the above result indicates that all the material will be optimally used.

- *SMAs Subject to Bending Deformation.* Let us assume a bending element of rectangular cross section as depicted in Figure 3.9b. Where the element

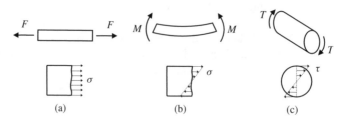

(a)   (b)   (c)

Figure 3.9 Typical load cases in SMA actuators: (a) pure tensile stress, (b) pure bending and (c) torsion.

is subject to pure bending, the normal stress at any cross section will vary linearly:

$$\sigma(x) = \frac{M_b x}{I} \tag{3.6}$$

where, M_b is the applied bending moment, x is the distance from the neutral plane and I is the moment of inertia, which for the rectangular cross section of the figure is $I = wh^3/12$.

For this particular load case, the maximum strain in the material will occur at the upper and lower limits of the cross section, and can readily be computed from Equation 3.6 considering that $\epsilon = \sigma/Y$. It can be shown quite simply that upon pure bending, the ratio of average to maximum stress in the material for this particular cross section is

$$\frac{\hat{\sigma}}{\sigma_{\max}} = \frac{1}{2} \tag{3.7}$$

Low values for this ratio indicate that the material is suboptimally used as an actuator. In fact, since the maximum strain at the cross section constitutes the recoverable load limit, a low ratio will indicate that most of the material is still far short of the strain limit. However, when the actuator is heated to the parent phase, all the material has to be heated, so that the resulting efficiency of the actuator will be less than in the tensile load case.

- *SMAs Subject to Torsion Deformation.* The subject in this case is the circular cross section in Figure 3.9c. Under pure torsion load of the wire, a shear strain state will develop across the section. The shear strain will not be uniform but will vary linearly from the neutral axis. The expression for the shear strain, $\tau(r)$, upon torsion loading is

$$\tau(r) = \frac{M_t r}{J} \tag{3.8}$$

where M_t is the torsion moment, r is the radial distance from the neutral axis and J is the polar moment of inertia of the cross section, which is $J = \pi d^4/8$ for this particular case.

Again, the ratio of average to maximum shear stress in the cross section can readily be demonstrated to be

$$\frac{\hat{\tau}}{\tau_{\max}} = \frac{2}{3} \tag{3.9}$$

In this case, the shear stress will be resisted by the material's shear modulus, and the relationship between shear strain and stress will be $\gamma = 8\tau/3Y$. The upper limit for the actuator load will be imposed by the maximum shear strain. Moreover, this load case is more sensitive to stress by a factor of 8/3, and, therefore, the torsion factor in Equation 3.9 must be corrected to yield

$$\left[\frac{\hat{\tau}}{\tau_{max}}\right]_{corrected} = \frac{1}{4} \tag{3.10}$$

The optimality in using the material for actuator applications for each load case is closely related to the load factors in Equations 3.5, 3.7 and 3.10. The efficiency of the actuator, as the ratio of output mechanical work to input heat, will be highly dependent on efficient use of the material.

Where the one-way shape memory effect is used in combination with external biasing concepts to develop shape memory actuators, cyclic operations can be treated like the operation of a heat engine (Thrasher *et al.* (1994)). Maximum theoretical efficiency in heat engines is determined by the high- and low-temperature reservoirs of the Carnot cycle according to the expression

$$\mu_{Theoretical} = 1 - \frac{T_L}{T_H} \tag{3.11}$$

In Equation 3.11, the low temperature corresponds to M_f, while the high temperature corresponds to A_f. With typical values of transformation temperatures and transformation temperature span as described in Section 3.1.1, the maximum theoretical efficiency is of the order of 10%. The theoretical maximum efficiency increases with the temperature of the reservoirs. Therefore, higher transformation temperatures are beneficial in terms of efficiency of the actuation process.

Under real conditions, one might expect practical efficiency to be rather low. Thrasher *et al.* (1994), who conducted a thorough analysis for tensile loading of SMAs wires, reported practical efficiency of the order of 3%. When the analysis is extended to pure bending and torsion loads, the figures are even lower, as one would expect from the foregoing analysis.

Shape recovery concepts

In general, there are three common concepts for implementing shape memory actuators, namely, extension wires, benders and helicoidal springs. These make use of the three load cases described in the previous section.

Extension wires are most appropriate when high forces at low displacements are required. Even though strain recovery up to 8% can be achieved with extension wires, in practice this value is restricted to 3–4%, which results in improved fatigue behavior.

Helicoidal springs are chosen wherever large deformations at relatively low forces are required. Almost any required displacement can be met by the proper choice of spring dimensions. Because they are less efficient than extension wires, helicoidal springs are most commonly used in passive applications.

Shape memory benders provide moderate displacements at very low forces. The displacement in shape memory benders is limited in practice by the maximum strain at the outermost edge of the cross section. If the strain limit is 3–4%, the ratio of bending radius to wire radius must be limited to approximately 30–25. In practice, this means that bending a shape memory wire 1 mm in diameter around a cylinder 30 mm in radius will produce a maximum strain of 3% in the wire.

Bias force concepts

If the one-way shape memory effect is to be used in the implementation of shape memory actuators, a restoring force must be applied to deform the alloy upon cooling to the α-phase. There are three main biasing concepts: constant bias load, variable bias force and imposed bias displacement.

- *Constant Bias Load.* This biasing concept is practically implemented either by using the weight of a suspended mass to restore the actuator to its deformed position or by means of constant force springs. See the inset in Figure 3.10 for a schematic depiction.

  The effect of the constant bias force can be clearly seen in stress versus strain and stress versus temperature graphs (see Figure 3.10 for the stress–strain representation). In the stress–strain graph, the actuator in austenite phase (B) is deformed upon cooling to the martensite phase (A) under the effect of the constant load Mg.

  In the stress–temperature graph, the actuator is again deformed under a constant load. Since the bias force is constant, there is no shift in the transformation temperatures at any point in the actuation–deformation cycle.

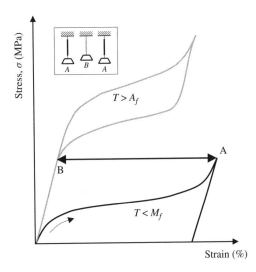

Figure 3.10   Constant bias load: the stress is kept constant throughout the actuation cycle.

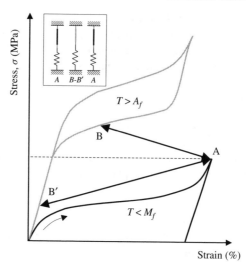

Figure 3.11   Variable bias load: the stress is altered (either increased, B, or decreased, B') while loading the SMA.

- *Variable Bias Force.* In most practical applications, the restoring force is supplied by a bias spring. The load being exerted on the actuator will vary as it becomes deformed. For a schematic representation of this biasing concept, see the inset in Figure 3.11.

  In the case of variable bias force, the loading trajectory, in both the stress–strain and the stress–temperature graphs, depends on the particular configuration and geometry of the biasing concept. In the most straightforward implementation, actuator and spring are collinear so that the applied bias load will decrease as the actuator is deformed from the austenite (B) to the martensite phase (A) (see curve B–A in Figure 3.11).

  However, the reverse bias concept (see Mertmann *et al.* (2002)) can also be implemented. In this case, the bias load increases as the actuator is deformed from austenite (B') to martensite (A) (see curve B'–A in Figure 3.11).

  Both the variable bias concepts have implications for the available stroke and for the transformation temperatures. On the one hand, reverse variable bias force produces a larger actuator stroke. On the other hand, the increased bias force resulting from the reverse bias concept causes a shift of M_f to higher values, which in turn may lead to improved performance upon cooling.

- *Imposed Bias Displacement.* In the two previous bias concepts, an external element imposes a force on the actuator, and this leads to a deformation upon cooling. On heating, the actuator imposes the displacement on the load.

  The bias concept can be implemented by using two antagonistic shape memory actuators. Each actuator is deformed by means of its counterpart. This

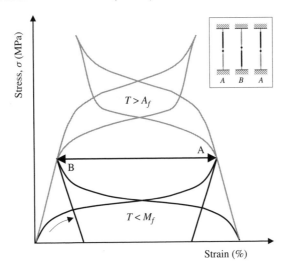

Figure 3.12 Imposed bias displacement: antagonistic SMA actuator imposes a displacement on its counterpart.

approach is conceptually different from the previous ones in that the restoring element (the second actuator) imposes a displacement rather than a force. This biasing approach is schematically depicted in the inset in Figure 3.12.

3.2.2 Material considerations

Nitinol (Ni50–Ti50) is the most commonly applied shape memory alloy. Small changes in the relative composition of both elements produce significant changes in the alloy properties. Both the hysteresis and the overall transformation temperature span differ slightly for different NiTi alloys. In addition, alloying can significantly affect the transformation hysteresis.

Nearly all applications of the shape memory effect are based on nitinol. Only approximately 10% of these applications make use of NiTi alloys with added ternary or quaternary elements. The average electrical, mechanical and chemical properties of shape memory alloys are summarized in Table 3.1.

The addition of ternary elements in the composition of the alloys may be beneficial for some applications. The following paragraphs summarize the use of ternary elements (see Mertmann and Wuttig (2004)):

1. Copper, Cu, additions have been shown to reduce hysteresis to about 10 to 15 °C. Cu as a ternary element can be found in concentrations between 3 and 25%, although for practical workability, the maximum addition is limited to 10%. Cu has been shown to reduce the sensitivity of the overall alloy properties to variation in the concentration of Ni. Also, another effect

Table 3.1   Some application-specific properties of selected shape memory alloys.

Property	Units	Ni–Ti	Cu–Zn–Al	Cu–Al–Ni
Physical properties				
Melting point	K	1513–1583	1223–1293	1273–1323
Density	Kgm^{-3}	6400–6500	7800–8000	7100–7200
Thermal conductivity A, (M)	W/mK	18 (8.6)	120 (−)	75 (−)
Thermal coefficient A, (M)	10^{-6}/K	11 (6.6)	16–18 (−)	16–18 (−)
Specific heat	J/KgK	470–620	390	400–480
Transformation enthalpy	kJ/Kg	3.2–12.0	7.0–9.0	7.0–9.0
Corrosion	–	≈ Inox	≈ A_l	≈ A_l
Electromagnetic properties				
Resistivity	10^{-6} Ωm	1.0	0.07	0.1
Magnetic permittivity	–	1.002	–	–
Mechanical properties				
Young's modulus	GPa	70–98	70–100	80–100
Yield stress A, (M)	MPa × 10^2	2–8 (1.5–3)	1.5–3 (−)	1.5–3
Ultimate yield stress	GPa	0.700–2.0	0.7–0.8	1.0–1.2
Maximum deformation	%	40–50	10–15	8–10
Fatigue limit (10^6 cycles)	MPa	350	270	350
SME properties				
Transformation temperature, (A_s)	K	73–373	73–393	73–443
Hysteresis	K	20–30	10–20	20–30
Maximum deformation (OWE)	%	8	5	6
Maximum deformation (TWE)	%	2	0.8	0.8
Intrinsic damping	%	15	30	10
Maximum superelastic deformation, SC (PC)	%	10 (4)	10 (2)	10 (2)
Stress gradient	MPa/K	4–20	2–5	–

of Cu additions that is useful for applications is to reduce sensitivity of the transformation properties to transformation cycles.

2. Niobium, Nb, additions can expand hysteresis to over 100 °C. Expanded transformation hysteresis can be useful in some shape memory applications, particularly pipe couplings.

3. Platinum, Pt, or Palladium, Pd, can be used to increase transformation temperatures. However, these elements are not used in practice owing to the high concentrations that are required to produce significant increments in transformation temperatures and to their high cost as precious metals.

Cold working and heat treatment have less dramatic, but still measurable, effects on transformation hysteresis.

3.2.3 Thermal considerations

Proper driving of shape memory alloy actuators is based on thermal control of the shape memory alloy (see Section 3.3). The underlying idea is, by thermally controlling the alloy, to take the phase transformation from the martensite to the parent phase up to the precise point at which the recovered strain produces the required displacement of the drive.

Because the shape memory actuator is subject to a bias force, precise position control can be implemented in both directions upon controlled cooling.

The heating process is generally electrically driven; in this approach, the drawn current heats up the shape memory alloy. The power, P_h, electrically delivered to the actuator by *Joule* heating is

$$P_h = i_h^2 R \qquad (3.12)$$

where i_h is the electrical current drawn for heating and R is the electrical resistance of the shape memory actuator.

The thermal inertia of the shape memory actuator limits the time response of the actuator. In particular, the cooling process is the dominant factor limiting this response. If one looks at the thermodynamic equation describing the energy equilibrium during the martensitic transformation, one can draw several conclusions relating to the thermal optimization of shape memory alloy actuators. In general, the energy transformation in the alloy can be described by the following differential equation:

$$mC_p \frac{dT(t)}{dt} + m \Delta H \frac{d\xi}{dt} = hA(T_\infty - T(t)) + i_h^2 R \qquad (3.13)$$

where C_p is the specific heat of the alloy, m and A are the mass and external area of the actuator (the one effectively contributing to heat convection into the environment), h is the convection coefficient, ΔH is the latent heat for the transformation, ξ is the fraction concentration of martensite in the alloy at a given instant and T_∞ is the room temperature.

A close examination of Equation 3.13 will yield the following conclusions:

- *Actuation Cycle: Heating Process.* In the heating process, the dominant factor in Equation 3.13 is $i_h^2 R$. The power delivered to the actuator through Joule heating is used to raise the temperature of the actuator, $mC_p(dT(t))/dt$, and to drive the transformation from martensite to the parent phase, $m \Delta H (d\xi/dt)$.

  In principle, the actuation cycle linked to the heating process can be made as fast as is required, provided that enough power is available and that care is taken to avoid thermal damage to the actuator.

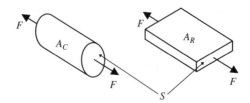

Figure 3.13   Difference in heat convection area for circular, A_C, and rectangular, A_R, cross sections: where the cross section, S, is the same, $A_R > A_C$.

- *Cooling Limits the Time Response.* The dominant factor upon cooling is $hA(T_\infty - T(t))$. The heat dissipated through convection is used to reduce the temperature of the actuator and to promote the transformation from parent phase to martensite. This is an exothermal transformation, and, consequently, the requirements for heat transmission to the environment are even more stringent.

- *Modification of the Convection Coefficient, h.* The cooling process is the dominant factor for the time response of the actuator, and so one means of reducing this response time is to increase the heat convection coefficient. Forced air cooling or fluid refrigeration can be considered for this purpose.

- *Maximization of the Effective Heat Transmission Area, A.* As a rule, the force requirements of the application will dictate the cross-sectional area, S, for wire actuators (see Figure 3.13). The effective heat transmission area is a function of the perimeter of the cross-sectional area. The ratio of perimeter to cross-sectional area increases if the cross sections are rectangular rather than circular. In this way, the heat transmission can be maximized for a constant available force, that is, a constant cross-sectional area.

- *Shift in the Position of Transformation Temperatures.* Another factor favoring faster cooling is the increased thermal gap between room and working temperature, $T_\infty - T(t)$. This can be achieved by selecting the appropriate shape memory alloy and the appropriate ternary constituent elements. See Section 3.2.2 for more details.

In addition to these design tips, following the thermal analysis of the transformation, some authors (Abadie *et al.* (1999)) have proposed that the *Peltier effect* be used for fast cooling of shape memory actuators.

3.3   Control of SMAs

3.3.1   Electrical heating

In the control of shape memory actuators, the shape recovery is thermally triggered. Depending on the type of application, the heating process may be either active or passive. In passive applications, usually called *thermal actuators*, the actuator is made to respond to increased temperature in its surroundings, for example, fire alarms in buildings or shape memory–driven water valves. See Case Study 3.2 for an example of thermal actuator applications.

In active applications, two different heating procedures are typically implemented, electrical heating and inductive heating. Inductive heating has proven to cause uneven heating of shape memory actuators, and this in turn leads to microstructural stress in the material. Consequently, electrical Joule heating is commonly used to drive SMAs.

The electrical equivalent of a SMA actuator load is a variable resistor (see Figure 3.14). The resistance of the electrical load will be subject to alterations as a result of changes in the geometry and electrical resistivity of the SMA actuator (see next section for more details).

As noted earlier, thermal hysteresis is inherent in shape memory actuation. This is clearly appreciable in Figure 3.15, which shows the hysteresis cycle in shape memory actuation when electrical heating and air convection cooling are used.

The figure describes the relationship between the heating current drawn and the resulting strain recovery. The experiment was conducted under a constant external load, which served the purpose of restoring the deformation upon cooling. As we will see in the coming sections, with strain versus current models it might be possible to use feed-forward control schemes to improve the actuator's performance.

3.3.2   Concomitant sensing and actuation with SMAs

In Section 3.1.1, we discussed the changes in electrical, mechanical and chemical properties of the alloys during the martensitic transformation. By tracking one of these properties, it is possible to follow the martensitic transformation, and, in principle, also all the characteristics associated to it, in particular, recovery of the actuator's shape upon heating.

The function of a SMA acting as a sensor is defined by its ability to track the shape change by monitoring the change in a different property. One of the electrical properties subject to change during the phase transformation is the electrical

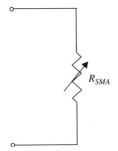

Figure 3.14   Electrical equivalent circuit for a SMA actuator load.

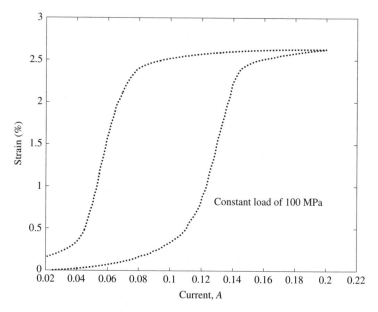

Figure 3.15   Strain versus temperature behavior of a shape memory actuator under a constant load of 100 MPa.

resistivity of the alloy. According to Hesselbach *et al.* (1994), the following can be said about the change in electrical resistivity in shape memory alloys subject to the martensitic transformation:

- The electrical resistivity of shape memory alloys, and, in particular, of $N_iT_iC_u$ alloys, is higher for the martensite phase than for the parent phase. The overall resistivity for the alloy at a temperature where both phases coexist can be estimated as a function of the fraction concentration of each phase:

$$\varrho = \xi \varrho_m + (1 - \xi)\varrho_a \qquad (3.14)$$

  where ξ is the fraction concentration of martensite in the alloy and ϱ_m and ϱ_a are the electrical resistivity for martensite and austenite respectively.

- The relationship between fraction concentration and deformation is linear.

In addition to the change in resistivity, the change in geometry in a wire actuator configuration (see Figure 3.16) contributes to the modification of overall electrical resistance.

The combination of these phenomena produces an approximately linear relationship between actuator displacement and electrical resistance. Moreover, the hysteresis in the electrical resistance versus deformation relationship will be very low if the shape memory alloys are well chosen.

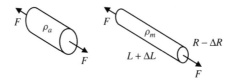

Figure 3.16   Geometry and electrical resistivity changes in SMA actuation.

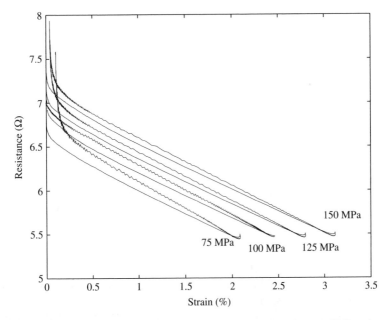

Figure 3.17   Electrical resistance versus strain relationship for a $N_iT_iC_u$ wire actuator as a function of applied load (the separation of the curves has been exaggerated
to stress this dependency).

Unfortunately, the electrical resistance versus strain relationship is also a function of the mechanical load (either the external load or the bias force used to
deform the alloy in the low-temperature phase). The resistance versus strain relationship for various different applied loads is shown in Figure 3.17. In this figure,
the different curves have been intentionally shifted apart to stress this dependency.
    On the basis of the electrical resistance versus strain relationship, the shape
memory actuator can be used in a concomitant way as a position sensor. This
exemplifies the mechatronic concept of combining different functions on the same
component. To do so, however, a model must be defined for the resistance–strain
relationship – that is, a sensor model.
    The simplest implementation for the sensor model can be defined by a curve
fitting the experimental data in Figure 3.17. Both linear and cubic polynomials

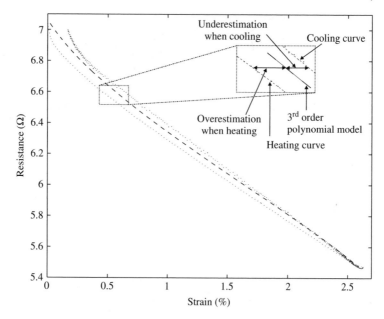

Figure 3.18   Third-order polynomial fitting of the resistance–strain relationship in a SMA. Detailed view of the fitting error due to the inherent hysteresis of the model.

have given good results in defining the sensor model. A closer look at the resistance–strain relationship for a particular load shows that a cubic polynomial fits the experimental data well enough (see Figure 3.18).

In spite of this good fit, the absence of pronounced hysteretic behavior in the resistance–strain relationship could cause either underestimation or overestimation of the actual position of the actuator. This is illustrated by the detailed view of the experimental curve and fitting polynomial (inset) in Figure 3.18. The lower dotted line represents the experimental data obtained when the actuator is heated and fully recovers the displacement imposed by the bias force. The upper dotted line represents the experimental data when it is cooled to the martensite phase, thus allowing the deformation induced by the bias force. The broken black line represents the polynomial that fits the experimental data.

3.3.3   Integration in control loops

It is difficult to control shape memory actuators precisely, mainly due to the intrinsic hysteresis and the inherent saturation of the shape memory effect. A number of control approaches for shape memory actuators are described in the literature. Proportional, integral and derivative (PID) control approaches have been widely explored. PID control strategies have been found to produce steady-state errors, to induce limit cycle problems and, most often, to cause position overshooting when heating and undershooting when cooling.

In coping with limit cycle problems, model-based feed-forward control approaches have been successfully applied in combination with PID control. However, exact models for the shape memory actuator are difficult to define and because of the thermal control of the actuators, they are always dependent on ambient conditions.

More elaborate control strategies have also been reported. Hasegawa and Majima (1998) used a hysteresis model for the actuator plant to compensate for the hysteretic behavior upon heating and cooling. Grant and Hayward (1997) used variable structure control to drive an antagonistic couple of shape memory actuators robustly and smoothly. Additionally, Song *et al.* (2003) reported the use of neural networks and sliding-mode robust controllers for precision tracking of shape memory actuators. And lastly, Lee and Lee (2004) adapted time delay control (TDC) schemes and PID control schemes based on TDC for accurate tracking operations.

As to the driving technique implemented in controlling shape memory actuators, both continuous control and modulation techniques have been used. It is reported (see Ma and Song (2003)), that pulse width modulation, PWM, helps reduce the energy consumption in electrical resistance actuation of shape memory actuators. This slight improves efficiency at the cost of reduced response time.

The remaining discussion in this section is an introduction to the way some control actions can be implemented on shape memory actuators. A few examples of traditional control approaches to drive shape memory actuators are described. First, the use of shape memory alloys as smart actuators, for example, concomitantly implementing sensing and driving actions, is illustrated. This is followed by a brief discussion of some basic control strategies, in this case PID and feed-forward control. The section closes with a description of a linearizing technique based on direct strain feedback (see Reynaerts (1995)) and a discussion of its effect on settling time and control performance.

The whole concept is illustrated in Figure 3.19. The black path in the control scheme corresponds to a classic PID control with the use of external position sensors. The linearizing concept is illustrated by the dashed line path. It is by means of this positive feedback scheme that the actuator plant is linearized. As a result, the controllers "see" a linearized plant (dashed line). Moreover, where the grey path is activated, an electrical current versus strain model of the actuator is used to implement a feed-forward control scheme. And finally, the smart actuator concept is implemented (dotted line path) wherever a resistance versus strain model of the actuator is used to estimate its position.

Smart actuation

The combination of sensing and driving functions on the same component leads to the what is called *smart actuation*. Section 3.3.2 introduced the resistance–strain models for shape memory actuators. These provide the basis for smart actuation.

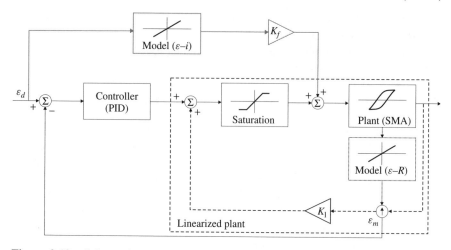

Figure 3.19   Schematic representation of the control concept: classic feedback control (black path), model-based feed-forward control (grey path), linearized plant concept (dashed path) and smart actuator implemented through a model-based estimation of the position (dotted path).

When these models are used to estimate the current position of a shape memory actuator, the resistance of the actuator is computed from the applied voltage and the drawn current. In order to avoid singularities in the calculation of the electrical resistance ($i = 0 \Rightarrow R \rightarrow \infty$), a minimum nonzero current is continuously applied to the actuator.

If the resistance versus strain model of the actuator is properly defined, it will allow correct estimation of the current strain (displacement), and so the actuator can also be used as a sensor. The control loop can then be closed on the basis of the position estimation.

Figure 3.20 shows the step response of a shape memory actuator under electrical resistance feedback. For reference, the current position based on an external sensor is also depicted. In this figure, the increase in strain is a result of the heating process.

It is apparent from the cooling part of the actuator position and the result of Figure 3.18 that the exact position is always either overestimated or underestimated. This depends on whether the actuator is being heated up or cooled down. In particular, the detailed view shows the transition point at which, after reaching the maximum undershoot in the actuator position, the electrical resistance estimation error becomes negative (the position is underestimated) after having been positive (the position was overestimated) during the cooling process.

As we discussed in previous paragraphs, resistance versus strain (displacement) models are only valid for a particular load. Then again, it might be interesting to assess the effect of using a particular model when varying loads are present. This situation is illustrated in Figure 3.21.

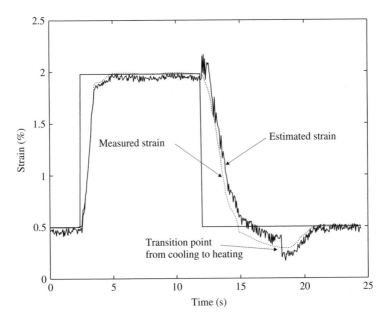

Figure 3.20   Step response of a shape memory actuator with electrical resistance feedback as an estimation of actual position.

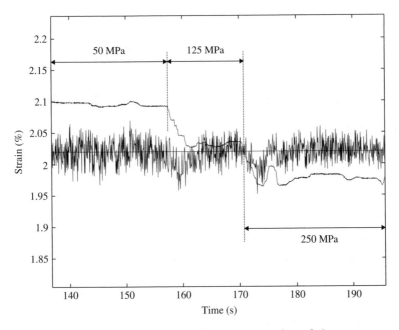

Figure 3.21   Effect of load changes on the smart actuation of shape memory actuators.

In this figure, the electrical resistance versus strain model for an actuator was developed at a particular load of 125 MPa. The noisy, high-frequency line represents the estimated position (strain) based on this model. The reference position is the horizontal line in the vicinity of 2% strain. In the central part of the graph (between 160 and 170 s approximately), the load on the actuator matches the model load, indicating a good estimation of the actual position.

For $t \leq 160$, a load of only 50 MPa was applied; then immediately after $t \geq$ 170, a load that was double the model reference load, 250 MPa, was exerted on the actuator. It is clear that even though the estimated position is always in the vicinity of the reference position (the control loop considers the estimation based on the electrical resistance as the right position), the real position differs from the reference.

However, even for such high load variations (more than 100% with respect to the model) the error due to the use of a wrong resistance–strain model is always less than 0.1%. For an actuator 100 mm in length, this would represent a position error of only 0.1 mm out of an overall stroke of about 4 mm.

Linearization techniques: Positive strain feedback

Positive strain feedback has been applied in the context of linearizing the actuator response in applications where accelerated cooling means, for example forced air convection or liquid cooling techniques, are available.

In this approach, a positive feedback loop is introduced, that is, the dashed path is activated (on the basis of either internal or external sensors). The loop gain must be kept low to ensure system stability. In addition, wherever the positive feedback loop gain is high, the system may become self-heated and cooling down is disabled.

Figure 3.22 shows a comparison of a PID control approach (a) on a shape memory actuator and (b) on a linearized version of the same actuator. In order to avoid secondary effects on the cooling performance of the positive strain feedback, the linearizing technique was only implemented during heating. It can be shown (see Figure 3.22b) that both the settling time and the overshoot performance of the system were greatly improved.

Feed-forward control of SMAs

Figure 3.15 introduced a quasistatic model for the strain versus temperature relationship in a shape memory actuator. This or any other type of actuator model can be used to predict an appropriate control action to drive the actuator to the reference position.

As noted earlier, shape memory actuator models are difficult to define and use, and they are always affected by the external ambient conditions and by the load on the drive. As a result, improvements of the actuator performance are most often insignificant compared to classic feedback strategies.

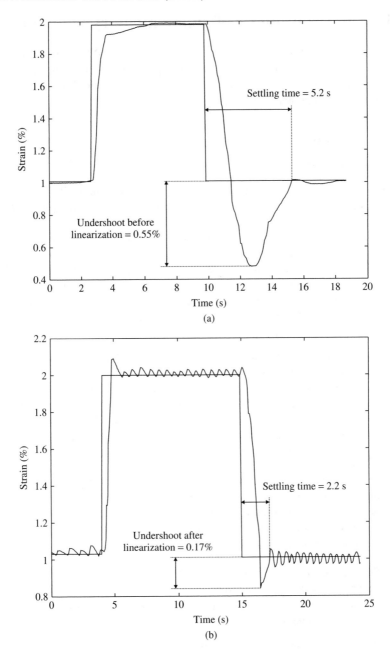

Figure 3.22   Effect of linearization (b) on the control performance of a SMA plant (a).

3.4   Figures of merit of shape memory actuators

This section summarizes the operational characteristics of SMA actuators and their behavior upon scaling. The first part outlines operational performance in terms of static, dynamic and environmental characteristics. The second discusses the effect of scaling on operational characteristics.

3.4.1   Operational ranges

Static performance

The main point of analysis in this category is the maximum available stroke and force. A distinction is made regarding the load case used to develop the actuator:

1. *Tensile wire.* The maximum recoverable strain that shape memory alloys can sustain is in the range 6–8%. The stroke of actuators developed on the basis of tensile wires is thus one order of magnitude larger than the stroke of piezoelectric and magnetostrictive actuators, also based on direct application of the piezoelectric and magnetostrictive effects respectively. The maximum stroke is usually limited by the number of actuation cycles required of the actuator, as these are subject to fatigue.

   The maximum available force depends on the cross-sectional area of the actuator. The maximum applicable stress producing recoverable strain is of the order of 150–250 MPa. This is the highest force density of all the emerging actuators described in this book, and, in particular, it outperforms pneumatic and hydraulic actuators. It is worth noting that this situation could alter in favor of the new carbon nanotube actuator for which the theoretical force, energy and power densities predict higher values.

2. *SMA bending actuators.* The stroke for bending actuators, as in the case of piezoelectric benders, is significatively larger than for tensile actuators. This is achieved at the expense of exhibiting virtually zero force. These actuators are therefore suitable for applications where a net displacement rather than a net work delivery is required, for example valves.

3. *SMA spring actuators.* Spring actuators provide the largest stroke of all SMA actuators. Moreover, if appropriately designed, spring actuators can meet virtually any stroke requirement. This technology also produces less force than tensile schemes.

   Actuators of this type are particularly suitable for developing the concept of thermal actuators; that is, they are best designed to provide some control action in response to environmental temperature changes rather than being included in motion control systems with resistive heating. This idea is further developed in Case Study 3.2.

Dynamic performance

The dynamic performance of SMA actuators is related to their energetic characteristics, their time response (and consequently frequency bandwidth) and to the efficiency of the actuation process. These aspects are reviewed in the following paragraphs.

- *Work Density Per Cycle.* This technology offers higher energy density per cycle than other emerging actuators. Theoretical considerations have led some authors to propose energy density values in the range $10-100$ J/cm^3 – see for instance Madou (1997).

  In the as-yet scarce commercial developments based on SMA technologies, the real energy density figures are in the range $10^{-1}-1$ J/cm^3 – that is, two orders of magnitude lower than theoretical predictions.

- *Time Response and Frequency Bandwidth.* SMA actuation is a slow process. It is limited by heat transport mechanisms during the cooling subcycle. It can be shown (Madou (1997)) that the time constant of such a heat transfer process, τ_H, is

$$\tau_H \propto \frac{\rho c_p r}{2h} \qquad (3.15)$$

  where ρ is the material's density, c_p is the specific heat, r is the radius of the SMA wire and h is the heat transfer coefficient.

  In practical applications, even though the time constant scales positively upon miniaturization, the maximum bandwidth is of the order of $1-5$ Hz (the high value for thin SMA films).

- *Power Density.* Power density is not commonly indicated in the literature. It can be worked out from the energetic density and the time constant or frequency bandwidth of the actuator (see previous item).

  Under ideal theoretical conditions of energy density and maximum bandwidth, the figures for power density are of the order of 10^2-10^3 W/cm^3. Where practical data are concerned, however, values can be as low as $10^{-3}-10^{-2}$ W/cm^3.

- *Energetic Efficiency.* The analysis of energetic efficiency is only relevant for resistance heated active applications of SMA actuators. In so-called thermal actuators, the system is designed to respond to changes in the ambient temperature. Since no additional input thermal energy is required, the efficiency cannot, strictly speaking, be defined.

  In the case of resistance heated active applications, the SMA actuator may be regarded, loosely, as a heat engine. At the usual transformation temperatures, the maximum theoretical efficiency predicted by a Carnot cycle is of the order of 10%.

In practice, the energetic efficiency is considerably lower and depends on both the transformation temperatures and the loading case. For tensile loading concepts, the practical limit for the efficiency is of the order of 3%, and less for bending and torsion load cases.

Other performance characteristics

Shape memory alloy actuators are particularly suitable for developing the concept of smart actuators, as they can be used concomitantly as sensor and driving technologies.

They can be integrated in functional or smart structures owing not only to the shape recovery process but also to their ability to change their stiffness and pseudoelastic behavior.

SMA actuators are suitable for incorporation in biomedical applications as they are biocompatible materials and are driven at low voltages, although heat dissipation could be a problem for these applications. Table 3.2 provides a summary of the various operational properties.

3.4.2　Scaling laws for SMA actuators

In this section, our analysis is restricted to resistance heated tensile actuators made with SMA wires. For other types of actuator geometry or for thermal actuators responding to changes in the ambient temperature, the reader is referred to Madou (1997) or Peirs (2001).

Table 3.2　Operational characteristics and scaling trends for SMA actuators.

Figures of merit	SMA wire	SMA bender	SMA spring
Force, F	High, ≤ 250 MPa	$\approx$ zero	low
Displacement, S	Up to 6–8%	High	High
Work density, W_V	Theory: $10-10^2$; Practice: $10^{-1}-1$ J/cm^3		
Power density, P_V	Theory: 10^2-10^3; Practice: $10^{-1}-1$ W/cm^3		
Bandwidth, f	Up to 5 Hz		
Efficiency, η	Theory limit 10%; Practice: $\leq 3\%$		
Scaling trends			
Force	$F \propto L^2$		
Stroke	$S \propto L$		
Work per cycle	$W \propto L^3$		
Energy density	$W_V \propto L^0$		
Bandwidth	$f \propto L^{-2}$		
Power density	$P_V \propto L^{-2}$		

As in most of the actuators described in this book, the force in SMA wire actuators is proportional to the cross-sectional area and thus scales as $F \propto L^2$, where L is the representative dimension of the actuator.

The stroke of the SMA actuator is proportional to its length and so can be said to scale as $S \propto L$. The work of the actuators is determined by the product of force and displacement, and, thus, the energy per cycle scales as $W \propto L^3$.

Here, we focus on the work density in order to assess how the previous figures of merit would evolve upon scaling. Since both mass and volume scale as $V \propto L^3$, the energy density or the specific energy density both scale as $W_V \propto L^0$. Therefore, no change in the energy density of the actuator is to be expected upon miniaturization.

SMA actuators are thermal actuators whose time response is limited by thermal transport mechanisms. In resistance heated active implementations, the limiting process is the cooling subcycle. Let us recall here the expression for the time response of a thermal actuator upon cooling:

$$\tau_H \propto \frac{\rho c_p r}{2h}$$

where the various different symbols have been previously introduced.

If we assume that both the specific heat, c_p, and the density of the material, ρ, remain constant upon scaling, which is a reasonable assumption, the time constant of the actuators scales as

$$\tau \propto \frac{L}{h} \tag{3.16}$$

It has been reported (see Peirs (2001)) that for small actuators ($L \leq 10$ mm), the heat transfer coefficient upon free convection (which applies in most cases) scales as L^{-1}. Consequently, the time constant of the actuators scales as $\tau \propto L^2$, and the frequency bandwidth scales as the inverse of the time constant, $f \propto L^{-2}$.

Since the power density can be expressed as the product of the energy density and the frequency, it scales as $P_V \propto L^{-2}$.

From all the above conclusions, we may expect SMA actuators to perform better upon miniaturization, since the specific power, time response and force density would scale positively upon size reduction. See Table 3.2 for a summary of the scaling trends of SMA actuators.

3.5 Applications

Case Study 3.1: Shape memory actuated latch mechanism for precision equipment during launch of spacecrafts

Case Study 2.5 introduced the application of stacked piezoelectric actuators as driving elements of an XYZ precision-scanning device for an atomic force microscope aboard the European Space Agency's ROSETTA–MIDAS instrument (Ariane 5 mission). The full design, manufacturing and testing under four different test models (two qualification models EQM and QM, and two flight models FM and FSM) were undertaken by Cedrat Technologies (Meylan, France).

Because of the high vibration and shock levels sustained by equipment aboard spacecrafts during launch, the scanning stage must be locked during launching. This section describes the development of a latching system with two degrees of freedom for the XY phase of the scanning stage. The novel design and implementation makes use of two shape memory actuators from TiNi Aerospace, the Frangibolt® FC2-16-31SR2 actuator. The text and illustrations in this section are courtesy of Cedrat Technologies (Meylan, France).

The basic principle of the latch mechanism is the use of a fastened compression spring to release the latching mechanism. The spring is initially prestressed by means of a notched fastener. The high force density and high available recoverable strain of a shape memory actuator is used to break the notched fastener restraining the spring-actuated release mechanism.

The system further includes a latch status indicator. The status indicator comprises two gold contacts that form a closed circuit while the latch mechanism is locked. When it is unlocked, the status indicator will show an open circuit condition.

Figure 3.23 shows a schematic view of the working principle of the latching mechanism. The fixed frames ⓐ and ⓑ and the moving frame ⓒ are initially

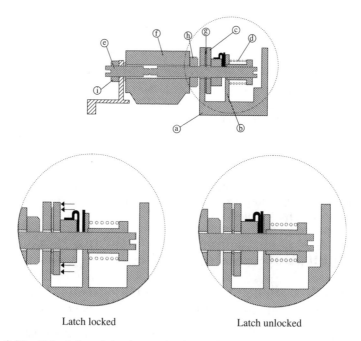

Latch locked   Latch unlocked

Figure 3.23 Principle of latch mechanism (Courtesy of R. Le Letty and F. Claeyssen. Reproduced by permission of CEDRAT TECHNOLOGIES).

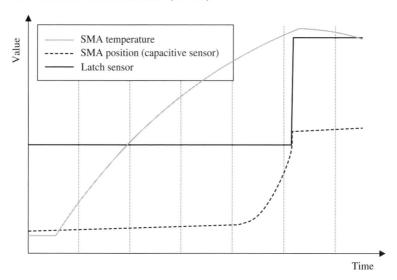

Figure 3.24   Chronogram of worst-case ($-20\,^\circ$C) latch operation (Courtesy of R. Le Letty and F. Claeyssen. Reproduced by permission of CEDRAT TECHNOLOGIES).

prestressed, together with the return spring ⓓ in the locked condition. The initial prestress is maintained by a notched fastener, ⓔ.

The latching process is thus accomplished by the friction planes pressing the friction washer ⓖ. The notched fastener, ⓔ, and the SMA actuator, ⓕ, are locked together by nuts ⓗ and ⓘ. Once the SMA actuator is activated, it will stretch the fastener. Since the recoverable strain is much higher than the yield strain of the fastener, the SMA action will first cause plastic deformation and eventually failure of the notched fastener.

The chronogram (Figure 3.24) shows the worst-case ($-20\,^\circ$C) condition of operation for the latch mechanism unlocking procedure. The SMA actuator is driven at a constant voltage between 22 and 36 V. The chronogram shows the temperature rise due to electrical heating, the latch status indicator and the elastic–plastic deformation of the notched fastener effected by a capacitive sensor. The latch mechanism unlocking cycle takes approximately 220 s to complete.

Case Study 3.2: Automatic oil-level adjustment in high-speed trains

This section describes the application of a shape memory actuator–driven oil-level adjustment device implemented in the Shinkansen bullet train in Japan. The oil-level-adjusting device described in this section is being used in the latest models of the Shinkansen Nozomi-500 and Nozomi-700 trains. The information and figures in this section are a courtesy of K. Otsuka and T. Kakeshita and were originally provided by Central Japan Railway Co. and Toyo Denki Seizo KK.

The principle exploited in this application case is the use of shape memory actuators as thermal actuators. In this exploitation, there is a temperature-sensitive SMA spring and a temperature-insensitive bias spring, which are mounted in series. In this application, then, the concept of a variable biasing force is implemented as explained in Section 3.2.1.

The oil-level-adjusting system is used in the Shinkansen train's driving-gear unit (Otsuka and Kakeshita (2002)). The driving-gear unit consists of a pinion and a gear, partially submersed in oil for lubrication (see Figure 3.25a and b). The oil temperature must be kept low when the train accelerates to high speed. The increase in oil temperature is determined by the turbulence level in the oil, and, therefore, the oil level must be kept low in Room A, as shown in Figure 3.25c.

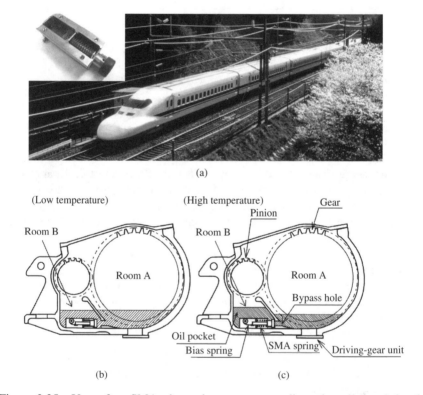

Figure 3.25 Use of a SMA thermal actuator to adjust the oil level in the Shinkansen Nozomi-700 bullet train. (a) Thermal actuator (inset) and Nozomi-700 train; (b) bias spring compressing the SMA while at low temperature and (c) SMA-driven valve in closed position at high temperature (Reproduced by permission of MRS Bulletin).

To achieve oil-level control, the driving-gear unit is divided into two rooms, A and B. Both compartments are connected by a bypass hole. At low temperature, a higher oil level may be allowed in room A (see Figure 3.25b). At high temperature, the oil level must be low (see Figure 3.25c).

The SMA-based control valve is shown inset in Figure 3.25a. It consists of a SMA compression spring and a bias belt–type spring. When the oil temperature is low, the bias spring compresses the SMA thermal actuator to its open position. In this condition, compartments A and B are connected and a high level of oil is allowed in room A. When the temperature increases, the thermal SMA actuator closes the bypass hole, and the gear and pinion rotate to partially evacuate oil from room A to room B. This reduces the oil turbulence and, hence, also the temperature.

The system described in this application case and shown in Figure 3.25 can maintain the maximum oil temperature at about 80 °C as compared to the equilibrium temperature of about 110 °C without the oil-level-adjusting SMA system.

Case Study 3.3: SMA-driven active endoscope with resistance feedback

There is constantly growing interest among the robotics scientific community in the application of robotics technologies to minimally invasive surgery. A number of prototypes have been reported in the literature during the last two decades. Minimally invasive surgery imposes strict requirements on devices, and, particularly, on the actuator technology.

Adaptations of more traditional technologies (microhydraulic cylinders) have been proposed. What interests us here is the application of shape memory alloy actuators in the field of surgery. The first reported application of SMA actuators in the field of active endoscopic tools for minimally invasive surgery was by Ikuta *et al.* (1988).

Endoscopes are used in colonoscopy and other techniques to allow diagnosis through imaging or biopsy. They are usually equipped with conduit wires so that they can be inserted in the colon and stomach. The process of guiding the endoscope is difficult both for the physician and the patient.

During the early 1980s, the team led by Prof. Hirose at Tokyo Institute of Technology developed a prototype of a cable-driven serpentine robot. The prototype allowed the operator to remotely actuate independent bending segments so that the shape of the so-called "ELASTOR" could be adapted to the task.

The next step in the evolution toward a minimally invasive tool was achieved by implementing shape memory actuators to actively bend the different segments in the active serpentine.

SMA actuators exhibit relatively low electrical resistance, which is utilized through Joule heating to induce the phase transformation. Prof. Hirose's team introduced a novel configuration of the various different SMA actuators by placing them mechanically in parallel and electrically in series. In this way, the electrical

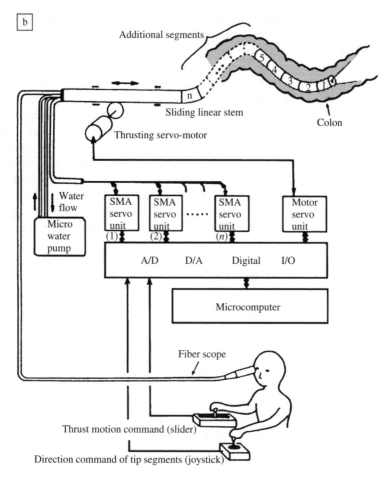

Figure 3.26   Block diagram of the control architecture for the SMA-driven active endoscope, (Reproduced by permission of MRS Bulletin).

resistance of the actuators can be increased so that a higher-voltage, low-current power source can be used to drive the serpentine robot.

In controlling the SMA-driven active endoscope, use is made of the intrinsic sensor capability of this technology. The electrical resistance of the different SMA wires is fed back to the controller for optimum driving, in a very compact solution. The servo-control system is shown as a block diagram in Figure 3.26.

Additionally, the SMA-driven serpentine endoscope is mounted on a linear-displacement servo drive. In this way, the endoscope can be axially positioned in addition to being adaptable in shape.

The active endoscope prototype has an overall diameter of just 13 mm and a total length of 250 mm. Five active segments are implemented. The controller

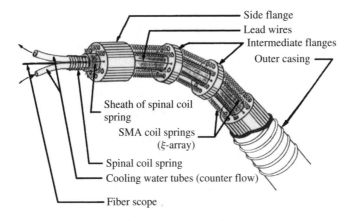

Side flange
Lead wires
Intermediate flanges
Outer casing

Sheath of spinal coil
spring
SMA coil springs
(ξ-array)
Spinal coil spring
Cooling water tubes (counter flow)
Fiber scope

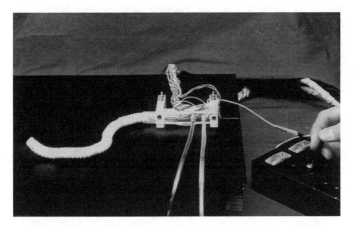

Figure 3.27   Schematic configuration of the SMA-driven active endoscope and picture of the final prototype, (Reproduced by permission of MRS Bulletin).

attains the reference bending angle with respect to the first segment and synchronously drives the following segments at the insertion speed determined by the linear-displacement servo drive, thus ensuring a smooth operation. The final configuration of the system can be seen in Figure 3.27.

Case Study 3.4:   SMA-driven implantable device for liquid drug delivery

Many medical treatments rely on the long-term administration of drugs. Some of these treatments require localized administration of drugs several times a day over long periods. Pathologic neurological conditions normally provoke motor disorders (tremor, spasticity, dyskinesia, etc.) and painful episodes. This type of medication is particularly common in neurological disorders.

In the particular case of spasticity, intrathecal baclofen therapy (IBT) has proven to be effective. Here, an implantable intraspinal drug infusion device is used. This consists of an implanted drug pump surgically placed under the skin, an implanted catheter and an external programmer. The pump stores prescribed amounts of baclofen and releases them into the intrathecal space of the spinal cord.

Patients return to the clinic for regularly scheduled pump refilling and programming. Refill schedules vary from once a month to every three months, depending on the amount of medication required by the patient. The pump is refilled by inserting a needle through the patient's skin into the drug reservoir in the pump. The intrathecal administration of baclofen results in longer reductions of spasticity and improved quality of life, and patients generally experience a decrease in the number and intensity of painful episodes. This type of intervention is also suitable for other disease conditions (diabetes, pain treatment, chemotherapy, etc.).

This case study describes the design of an implantable device for controlled drug delivery based on shape memory alloy actuators. The prototype was developed at the Production, Machine Design and Automation Division of the Department of Mechanical Engineering at the Katholieke Universiteit Leuven (PMA-KULeuven). The pictures and information for this case study are courtesy of J. Peirs.

The prototype of the drug delivery system includes a multidose reservoir, a single-dose reservoir, valves for controlled drug delivery, tubing and an electronic controller, including an antenna for inductive transcutaneous power supply. A schematic cross section of the drug delivery device is shown in Figure 3.28.

It is extremely difficult to control the dosing of drugs by means of a proportionally controllable miniaturized valve, and so the dosing strategy consists in two

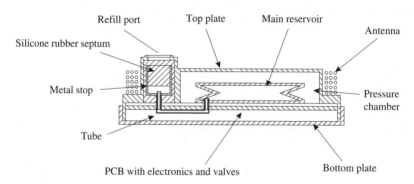

Figure 3.28   Cross section of the drug delivery system (Courtesy of Jan Peirs. Reproduced by permission of PMA-KULeuven).

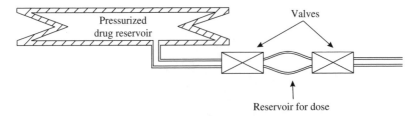

Figure 3.29   Concept for the drug delivery system (Courtesy of Jan Peirs. Reproduced by permission of PMA-KULeuven).

reservoirs as depicted in Figure 3.29. A pressurized multidose reservoir is used to hold medication for long periods (around 40 doses) and to refill a small reservoir for a single dose.

When the left valve in Figure 3.29 is opened, the drug flows from the multidose reservoir into the small reservoir. When refilling is complete, the left valve closes and the right valve opens to administer the drug to the body. This dosing strategy makes for increased safety, since in the event that a valve breaks, the second valve ensures that there is no uncontrolled drug delivery to the body. On the other hand, it is a fixed dosing system: in other words, the dose can only be changed by repeated cycling.

The essential part of the drug delivery system is the valve. The SMA-driven valve developed at PMA-KULeuven relies on a pincher placed on a silicone tube. When the valve is not actuated, the pincher presses the silicone tube and obstructs the flow of drug. The pincher can be opened by means of a SMA actuator.

The demands on the actuator are stringent for this type of application. A high force level is required. Electrostatic, electromagnetic and piezoelectric multimorph (bender) actuators do not deliver enough force to open the valve. On the other hand, piezoelectric stacks and magnetostrictive actuators provide too small a displacement. Shape memory alloy actuators therefore appear to be the most appropriate solution in this case.

The valve concept constitutes a trade-off between minimal dimensions, minimum number of parts (always difficult to manufacture on a small scale) and minimal energy consumption. In this concept, joints have been replaced by elastic hinges, and parts are joined by gluing, melting and welding.

The concept is shown in Figure 3.30: the valve is normally closed (safety) and only opens when the force generated by the SMA actuator deforms the elastic aluminum structure. When the SMA is not actuated, the elasticity of the valve structure closes the tube again. In this concept, a screw is used to pretension the SMA wire. The overall size of this first prototype is $15 \times 10 \times 2$ mm. The SMA actuator is a wire 120 µm in diameter with a A_s temperature of 65 °C (see Figure 3.30).

Figure 3.30   Aluminum valve prototype (Courtesy of Jan Peirs. Reproduced by permission of PMA-KULeuven).

Figure 3.31   Second, miniaturized prototype of a SMA-activated valve (Courtesy of Jan Peirs. Reproduced by permission of PMA-KULeuven).

A second, miniaturized prototype SMA-driven valve has been developed at PMA-KULeuven. For this second prototype, glass fiber–reinforced polyetherimide was used as the structural material. The high strength (yield stress 150 MPa) and Young's modulus (7.6 GPa) of this material ensures large forces in a small volume. The same SMA wire as in the previous prototype was implemented for this second case. The wire is used at a stress of 150 MPa and a strain of 3%. Figure 3.31 shows a photograph of the valve and Figure 3.32 shows a close-up detail of the tips of the valve while it is pressing the silicone tube.

Figure 3.32   Detailed view of the valve tip and pressing operation on the silicone tube (Courtesy of Jan Peirs. Reproduced by permission of PMA-KULeuven).

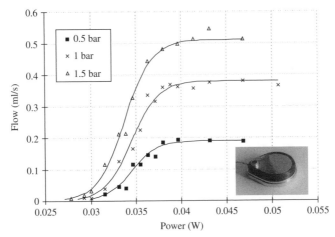

Figure 3.33   Flow characteristics of one of the SMA valves and (inset) view of the drug delivery device (Courtesy of Jan Peirs. Reproduced by permission of PMA-KULeuven).

The inset in Figure 3.33 shows the final prototype with the antenna used for inductive power supply (dark part forming the edge of the device). Figure 3.33 shows the flow characteristics of a single valve as a function of the power input to heat the SMA actuator and the pressure in the fluid (water at ambient temperature).

4

Electroactive polymer actuators (EAPs)

The family of electroactive polymer (EAP) actuators is a broad set of dissimilar technologies combining various different transduction phenomena, on different substrate materials and producing diverse actuation characteristics. Most of the phenomena underlying the actuation process are not fully understood, as some of the EAP technologies were only discovered in the last ten years.

If the trend observed in other actuator technologies (i.e. a long time lapse between the discovery of the transduction phenomena and its practical application) also prevails in these heterogeneous technologies, practical embodiments can only be expected after a number of decades.

This chapter, then, differs conceptually from the other chapters of the book in that it describes technologies that are evolving on a daily basis. The book therefore confined itself to introducing the basic principles of operation of EAP actuators and indicating trends in design and control issues.

A first section describes the actuation principles. As noted above, these differ from one technology to another and involve electrical charge reorientation, mass transport mechanisms, chemically triggered volume transitions, electrostriction and Maxwell forces.

The sections devoted to design and control indicate how particular aspects of these transduction phenomena are expected to impact design concepts and control strategies. In this connection, we elaborate on switching control as a means of coping with variable dynamics and volume-phase transitions in these technologies.

Finally, following an analysis of theoretical projections of performance, the chapter introduces a few case studies. These focus on the application of conducting polymer (CP) actuators as one of the technologies that are closest to practical application, with particular attention to the biomedical application domain.

Emerging Actuator Technologies: A Micromechatronic Approach   J. L. Pons
© 2005 John Wiley & Sons, Ltd

4.1   Principles

4.1.1   Wet EAP actuators

Responsive gels

A polymer gel is a structure of cross-linked polymers swollen in a solvent. The process of gelation starts from a colloidal suspension subject to either a physical or a chemical process. After gelation, part of the solvent is retained in the structure, giving rise to the characteristic properties of gels.

The actuation process in polymer gels is directly related to the ability of these structures to partially take up or expel the solvent in response to external stimuli. The change in volume associated with the expulsion or uptake of solvent provides the ability to drive external loads (see Figure 4.1).

The volume transition in polymer gels is triggered by changes in their *osmotic pressure*. The osmotic pressure tends to expand (positive pressure) or compress (negative pressure) the gel and is the product of intermolecular interactions, electrical charge and the pressure of moving particles in the gel. The main constituents of the osmotic pressure are as follows:

1. *Van der Waals force*, resulting from interactions between neighboring molecules in a nonpolar solvent.

2. *Hydrophobic interaction*, occurring between water molecules in the gel and hydrophobic polymer chains.

3. *Hydrogen bonding*, resulting from the interaction of hydrogen atoms located between two electronegative atoms.

4. *Electrostatic interactions*, taking place in the so-called polyelectrolyte, that is, with fixed charges on the network. Ions of opposite charge will be

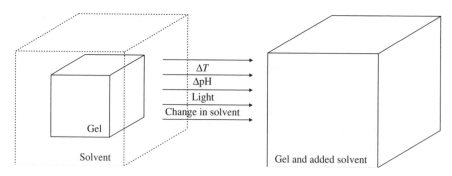

Figure 4.1   Volume change in polymer gel actuators as a consequence of solvent uptake in response to external stimuli.

attracted to provide electrical neutrality. The pressure produced by the relative concentration of ions will result in osmotic pressure. A particular instance of such an electrostatic interaction is the *hydrogen ion pressure.*

In addition to the osmotic pressure resulting from the above interactions, the *rubber elasticity* of the gel will contribute to the net force, tending to shrink or expand the gel. The rubber elasticity will always oppose changes in volume away from the equilibrium state.

According to Flory-Huggins' theory of gels, the osmotic pressure in the gel, Π, will comprise the following factors:

$$\Pi = \Pi_M + \Pi_{el} + \Pi_{ion} \tag{4.1}$$

where Π_M is due to the *polymer–polymer affinity* and comprises interactions 1 through 3 as described above, Π_{el} represents the elastic forces, that is, rubber elasticity, and Π_{ion} is the product of electrostatic interactions (interaction 4).

The fundamental parameter describing the pressure–volume relationship in polymer gels is the so-called polymer–solvent interaction parameter, χ. χ describes the solubility of the polymer in the solvent. For a polymer gel, small changes of χ about the equilibrium point will cause solvent to be expelled or taken up, and the gel will undergo volume changes accordingly.

There are a number of stimuli that can disturb the equilibrium between gel and solvent. Polymer gels have been reported to respond to temperature, pH, solvent composition, electric field and light. Table 4.1 summarizes some of the active polymer gels reported in the literature and gives details of activation stimuli and response time. These stimuli will modify the osmotic pressure and, thus, promote expansion or contraction.

One important factor in the dynamics of polymer gel actuation is the intrinsic diffusion-limited process of solvent expulsion or uptake. According to Fick's law of diffusion, the response time of the process, t_{gel}, will be limited by the characteristic distance in the gel, x and the diffusion coefficient, D:

$$t_{gel} = \frac{x^2}{2D} \tag{4.2}$$

As a direct consequence of the diffusion limit on the response time, the performance of polymer gel actuators is influenced to a large extent by the dimensions of the gel.

Ionic polymer metal composites (IPMC)

Ionic polymer metal composites, IPMCs, are also referred to as ionomeric polymer metal composites or ionic conducting polymer gel films, ICPFs. In the literature, they have been also referred to as 'soft actuators' or 'artificial muscles'. They are composite structures based on ion exchange membranes (Nafion, Flemion, Aciplex) and metal thin layers (platinum, gold). The ion exchange membranes

Table 4.1   Some responsive polymer gels. Activation stimuli and actuation characteristics.

Polymer gel	Stimulus	Response time	Actuation mode
Poly(acrylic Acid), PAA	pH	≈40 s	Discontinuous
Poly(vinyl alcohol), PVA–Poly(acrylic Acid), PAA	pH	Slow	–
Poly(acrylonitrile), PAN–Poly(pyrrole), PPY	pH	Slow	–
N-isopropylacrylamide, NIPA	Temperature	Thermally limited	Discontinuous
Poly(vinyl methyl ether), PVME	Temperature	≈1 s	–
Poly(acrylamide), PAM	Electric field	≈0.1 s	–
Poly(vinyl alcohol), PVA	Electric field	≈0.1 s	–
Poly(acrylamide), PAM	Solvent (Sodium Acrylate)	Slow	–
Poly(vinyl alcohol), PVA	Solvent (Water–Acetone)	Slow	–
Poly(acrylamide), PAM	Solvent (Water–Acetone)	Slow	–
N-isopropylacrylamide, NIPA	Visible light	Thermally limited	Discontinuous

usually employed in IPMC actuators have also been used in the context of fuel cell membranes.

The ion exchange membranes usually employed in IPMCs are perfluorinated ionomers. These are commercially available from several manufacturers; the most common compositions share the same backbone structure but comprise side groups of different nature and size.

The structure of IPMC actuators is shown schematically in Figure 4.2. The thickness of the ion exchange membrane is of the order of 200–300 μm, while that of the metal electrodes is of the order of 5–10 μm.

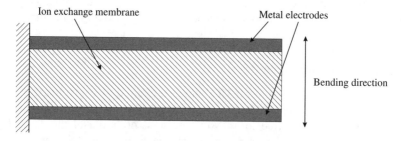

Figure 4.2   Schematic representation of an IPMC actuator.

The ion exchange membrane is a polyelectrolyte, that is, it has anions fixed to the structure. In the manufacturing process, counterions are introduced in the structure so that the overall electrical charge is balanced. The counterions in the structure can move freely, subject to electrostatic attractive and repulsive forces.

When a small electric field is applied between the electrodes, the counterions tend to diffuse to the corresponding electrode. The macroscopic result of this internal reallocation of electrical charge is a fast bending of the structure followed by a slow relaxation in the opposite direction (similar to the contraction–relaxation phenomenon in natural muscles (see Chapter 1)).

In IPMC structures, the fast application of a bending force causes a small voltage to be generated between the electrodes, so that IPM actuators can also be configured as soft sensors.

Of the various different EAP technologies, IPMCs are the ones that receive most attention from the research community. The main advantages of this as compared to other polymer technologies are as follows:

- Ion exchange membranes are readily available from several manufacturers: Nafion (DuPont), Aciplex (Asahi Chemical) and Flemion (Asahi Glass). IPMC actuators can be readily produced from these membranes and are easy to handle.

- They are driven at low voltage, typically 1 V and can therefore be actuated with standard power sources.

- They are mechanically stable and can also be used as sensors.

IPMC actuators also present some drawbacks. The main limitation resides in the bending actuation principle, which determines a force of virtually zero. An additional limitation is that IPMCs require humidity to work. They stop actuating if they get dry, so that for use, IPMC actuators have to be immersed in the solvent, with the attendant problem that this poses for packaging.

According to Nemat-Nasser and Thomas (2001), strategies for containing the solvent in the IPMC structure are being addressed as a means of limiting the restriction on use of external solvents.

As to actuation properties, IPMC actuators are subject to fast bending motion upon the application of low voltages. The time response for the first voltage-induced bending motion is of the order of a few milliseconds. This bending displacement is followed by relaxation, which is much slower, taking anything from several seconds up to minutes.

One of the limitations when using water as a solvent is that electrolysis occurs at driving voltages higher than 1.23 V. When electrolysis takes place, breakdown of water causes unwanted production of hydrogen and/or oxygen and raises the current requirements for IPMC actuators.

The work density or specific work of IPMC actuators is limited to 10 W/kg, according to estimations by Wax and Sands (1999). They have been proposed for use in biomedical applications where a displacement rather than a net force is required at low output power.

Conducting polymer actuators

Conducting polymer actuators are based on materials comprising a conjugated polymer backbone. Conjugated polymers are those whose structure presents alternating single and double bonds. Advantage is now being taken of the electrochemical characteristics of conjugated polymers and the structural properties of polymers to develop a new set of hybrid materials with optical, chemical or electrically triggered change in their mechanical and geometrical properties.

The most widely used conjugated polymer actuators are based on polypyrrole (PPy) and poly aniline (PANi), although other formulations based on poly thiophene (PTh) and poly paraphenylene vinylene (PPV) have also been proposed (see Sommer-Larsen and Kornbluh (2004)).

The mechanism of actuation is based on the uptake or expulsion of ions and solvent in the polymeric structure following an oxidation–reduction electrochemical process (Immerstrand et al. (2002)). The change in oxidation state causes a flux of charge from the polymer backbone, which triggers an answering flux of ions to electrically balance the charge. The corresponding mechanical response is a change in the stiffness properties of the conducting polymer and in its associated volume.

Two types of actuation processes can be defined in terms of how the conducting polymers are synthesized (see Sansiñena and Olazábal (2001)):

1. *The conducting polymer is grown in the presence of bulky anions.* The elements being taken up and expelled to compensate the electrical charge during the redox process are cations and solvent molecules. In this situation, the volume increases during the reduction process and decreases during oxidation. See Figure 4.3 for a schematic representation of the process.

2. *The conducting polymer is generated in the presence of small anions.* In this case, compensation of the electrical charge during the redox process is achieved through expulsion and uptake of anions and solvent molecules. The volume increases during the oxidation process and decreases during reduction.

The simple mechanisms above may be adapted to more complex behavior, with both anions and cations participating, as the polymer is cycled repeatedly (Bay et al. (2001)).

One of the limiting factors when using conducting polymers as actuators is the intrinsically slow diffusion process associated with expulsion and uptake (transport) of ions in the polymer structure. This makes it necessary to use thin layers of the conducting polymer if a fast response time is required. As a result, these actuators are more suitable for developing the micromuscle concept.

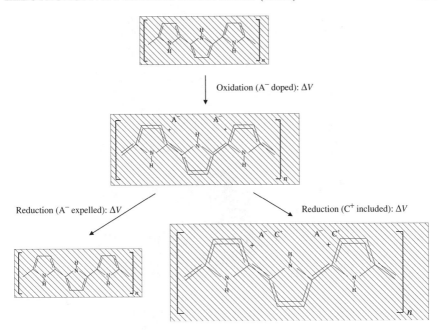

Figure 4.3   Schematic representation of oxidation–reduction processes causing volume change in CP actuators.

In particular, polypyrrole, one of the CPs most widely used in actuator applications, is usually combined with aqueous electrolytes in very thin layers, so that ion transport results in slow operation. See a PPy actuator grown on stainless steel in Figure 4.4.

The volume change in CP actuators is triggered by the application of an electric potential (typically of the order of $1-2$ V) between the polymer and the electrolyte. The volume change can be controlled by regulating the applied electric potential. The volume change associated with CP actuation is moderate compared to other EAP actuators and results in linear strain of the order of $2-5\%$. The strain rate (limited by diffusion) is of the order of $10\%/s$.

A mechanism commonly used to induce larger displacements (at the price of less available force) is the multimorph configuration principle. In this case, the active CP layer is laminated to an inactive supporting layer (Immerstrand et al. (2002)), usually referred to as a 'No Volume-Change Film' (NVC). In this configuration, the strain of the active CP film bonded to the inactive film causes a bending motion of the composite laminate film. The technique has also been analyzed in the context of several active layers alternatively laminated to inactive ones (Sansiñena and Olazábal (2001)).

The polymeric backbone structure provides good mechanical properties. The stiffness of the polymeric structure is high, of the order of $1-2$ GPa, and theoretical available stress for some polymers has been estimated at up to 450 MPa. However,

Figure 4.4   A 20-μm polypyrrole (PPy) actuator electrochemically grown on a 1-cm stainless steel electrode. Courtesy of S. Skaarup, DTU.

practical values ($\approx$ 34 MPa) still fall well short of the theoretical estimates (Madden (2004)).

The electromechanical coupling in CPs actuators is low (about 1%), making for poor efficiency. However, it has been reported that at steady state, CP actuators can sustain forces at virtually zero current consumption, thus constituting an effective blocking mechanism. Under these conditions, the actuator can block forces at a very low energy cost. See a PPy actuator on a testing set up in Figure 4.5.

Carbon nanotubes

Carbon nanotubes, CNTs, are the newest of all the transducing materials described in this book under the heading of electroactive polymer actuators. CNTs were first announced by Ijima (cited in de Heer (2004)), who reported the helical structure of carbon nanotubes and the coaxial arrangement in multiwalled nanotubes (MWNTs).

Carbon nanotubes come in two structural configurations, *single-walled nanotubes, SWNT* and *multiwalled nanotubes, MWNT.* MWNTs are obtained from nested SWNT structures, which in turn are developed from graphene sheets. Depending on the particular helicity indices or chiral vector, (n, m), CNTs may be either semiconducting or metallic (de Heer (2004)). If $(n - m)$ is a multiple of 3, the nanotube will have metallic properties.

The operating principle of carbon nanotubes as actuators is based on the use of CNT structures as electrodes in supercapacitors. The typical ionic charging process in conductive polymer actuators (involving either ion or counterion diffusion, so that response is limited by transport mechanisms) is substituted in CNTs by

Figure 4.5   A polypyrrole (PPy) actuator film (30 mm wide, 15 μm thick) mounted in a holder for the measurement of force, extension or stiffness. Courtesy of S. Skaarup, DTU.

electronic charge injection. CNTs are therefore not diffusion-limited, and a lower response time can be expected.

In a typical actuator configuration, the CNT structure (usually in the form of a paper sheet) is used as an electrode in an electrolyte (see Figure 4.6). Electronic injection of charge is achieved by the application of a voltage. The electronic charge is then compensated by electrolyte ions that accumulate on the outer face of the CNT structure to produce the so-called double layer (Baughman *et al.* (1999)).

The process whereby the charging of the electrode produces a strain comprises three factors:

1. *Quantum mechanical effects.* These effects are responsible for the expansion of the CNT structure upon injection of electrons and its contraction upon injection of holes.

2. *Quantum chemical effects.*

3. *Electrostatic double layer effects.* Both quantum chemical effects and double layer effects result in expansion of the CNT structure.

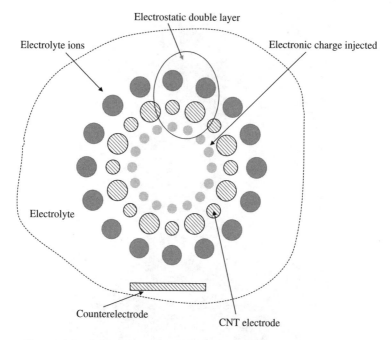

Figure 4.6   Principle of actuation with carbon nanotubes (CNT).

The predominant factor in the development of strain is the electrostatic double layer effect. CNTs combine a number of properties that make them a highly promising alternative for actuation. The strong electrostatic effect is due mainly to the large gravimetric surface area of these structures, the charge transfer properties of graphite, its electrical conductivity and impressive mechanical properties.

The volumetric work density per cycle, W_V, is defined according to the following expression:

$$W_V = \tfrac{1}{2} Y \varepsilon^2 \qquad (4.3)$$

where Y is the Young's modulus and ε is the strain. Likewise, the gravimetric work density per cycle, W_ρ, is defined as:

$$W_V = \tfrac{1}{2} Y \varepsilon^2 / \rho \qquad (4.4)$$

where ρ is the density.

It has been found that in MWNTs, Young's modulus depends on the radius, r_n, of the nanotube and that for $r_n < 5$ nm, Young's modulus is around 1 TPa. It has been estimated (Baughman *et al.* (1999)) that in SWNTs (with slightly lower Young's modulus than MWNTs) it may be possible to achieve a volumetric work density per cycle around 30 times higher than any other known actuator technology. If the gravimetric work density is taken into account, the figure rises to 150 times that of any other actuator technology.

In order for CNT actuators to attain these figures, the overall mechanical properties of bundled CNT actuators should be similar to the properties of single nanotubes (for which the previous estimates are predicted). In particular, the most critical factor resulting in different values for single and bundled structures is the gravimetric surface area, A_ρ. Although measured values of A_ρ for bundled structures are very high (of the order of 300 m^2/g), they still fall far short of the theoretical value for single tube structures, $\approx$1600 m^2/g for nanotubes 17 Å in diameter (see Baughman *et al.* (1999)).

4.1.2 Dry EAP actuators

We have discussed the principle of actuation for ionic (wet) EAPs in detail in previous sections. In this section, we focus on EAPs that respond directly to the application of an external electric field: *electric EAPs* or *dry EAPs*. In this case, mass transport no longer limits the response time of electric EAPs, and, thus, faster-responding polymers can be expected.

Electric EAPs can be further classified into *electrostrictive polymers* and *dielectric elastomers*. One feature they share is that stress and strain are quadratically dependent on the applied electric field. Here, we will examine the constitutive equations for both electric EAPs and highlight the similarities with piezoelectric actuators.

Although the functional relationship between strain and electric field is quadratic in the case of both electric EAPs, electrostrictive polymers are characterized by field- induced reorientation of electric dipoles, while dielectric elastomers rely on electrostatic interaction between electrical charges, commonly referred to as *Maxwell forces*.

In addition to response time, one of the main advantages of electric EAPs as compared to ionic EAPs is their chemical stability (Sommer-Larsen and Kornbluh (2004)). However, as we will see in the following paragraphs, stronger electric fields are required than in the case of electrically activated ionic EAPs.

Electrostrictive polymers

Electrostrictive polymer actuators are commonly based on ferroelectric polymeric materials, namely polyvinylidene fluoride (PVDF), nylon, polyurethanes and poly(vinylidene fluoride–trifluoroethylene) copolymer.

The resulting actuator properties in electrostrictive polymers are high actuation stroke (of the order of 5%), high energy density as compared to piezoelectric ceramics and moderate forces that render them suitable for diaphragm and bending actuators (Sommer-Larsen and Kornbluh (2004)).

The piezoelectricity and piezoelectric effect in the context of ceramic materials was introduced in detail in Chapter 2. The constitutive equations describing the electromechanical transduction in piezoelectric ceramics were developed in tensor notation and were compacted in matrix notation. For the sake of convenience, we recall here the constitutive equations for piezoelectric materials:

$$S_{ij} = c^E_{ijkl} T_{kl} + d_{mij} E_m$$

$$D_k = d_{kij} T_{ij} + \epsilon^T_{km} E_m \tag{4.5}$$

and in compact notation:

$$\left\{ \begin{matrix} S \\ D \end{matrix} \right\} = \begin{bmatrix} [s] & [d]^T \\ [d] & [\varepsilon] \end{bmatrix} \left\{ \begin{matrix} T \\ E \end{matrix} \right\} \tag{4.6}$$

The piezoelectric effect as described by Equation 4.5 is not present in all polymers. In particular, those polymers exhibiting inversion symmetry are not piezoelectric. However, all polymers exhibit a quadratic dependence of the strain on the polarization known as an *electrostrictive effect*.

The constitutive equation describing the quadratic relationship between strain and polarization is

$$S_{ij} = Q_{ijkl} P_k P_l \tag{4.7}$$

According to the equivalence between tensor notation and compact notation introduced in Table 2.1, the constitutive equation in compact notation is:

$$S_3 = Q_{33} P^2 \quad \text{and} \quad S_1 = Q_{13} P^2 \tag{4.8}$$

The electric state of a dielectric was described in Chapter 2 as a function of the electric field, $\vec{E}$, the electric displacement or charge density, $\vec{D}$, and the polarization, $\vec{P}$, according to the following expression:

$$\vec{D} = \epsilon_0 \vec{E} + \vec{P} \tag{4.9}$$

Equation 4.9 can be rewritten to obtain a direct relationship between applied electric field and polarization:

$$\vec{P} = (\epsilon - \epsilon_0)\vec{E} \tag{4.10}$$

Now, the relationship between strain and applied electric field is straightforward,

$$S_3 = M_{33} E^2 \quad \text{and} \quad S_1 = M_{13} E^2 \tag{4.11}$$

where $M_{33} = Q_{33}(\epsilon - \epsilon_0)$ and $M_{13} = Q_{13}(\epsilon - \epsilon_0)$.

Equations 4.11 can now be expanded in Taylor series around a DC bias electric field, E_0, for small variations of the electric field around the DC bias:

$$S_i = S_{i0} + \left.\frac{dS}{dE}\right|_{E=E_0} (E - E_0) + \frac{1}{2!} \left.\frac{d^2 S}{dE^2}\right|_{E=E_0} (E - E_0)^2 + \cdots \tag{4.12}$$

According to Equation 4.11, and given the small variations of E, the expansion of Equation 4.12 produces:

$$\Delta S_3 \approx 2M_{33}E_0\Delta E \quad \text{and} \quad \Delta S_1 \approx 2M_{13}E_0\Delta E \qquad (4.13)$$

The functional relationship between strain and applied electric field is quadratic. Equation 4.13 indicates that the electrostrictive behavior of small AC electric fields around a DC bias electric field may be regarded as the same as the piezoelectric behavior with a remanent polarization equivalent to the bias electric field, that is, $P_0 = E_0/(\epsilon - \epsilon_0)$.

With high electric fields, the real strain–electric field relationship usually deviates from the quadratic functional relationship described by Equation 4.11. The curve is said to exhibit saturation-like behavior. The situation is depicted qualitatively in Figure 4.7.

Dielectric elastomers

In soft polymeric materials, the electrostatic interaction between two electrodes with opposite electric charge can cause significant deformation. When electroded and subjected to a strong electric field, all dielectric polymers undergo electrostatic attraction in the thickness direction and electrostatic repulsion in the electrode plane (Sommer-Larsen and Kornbluh (2004)).

The forces resulting from electrostatic attraction squeeze the film in the thickness direction and expand it in a direction perpendicular to the applied field. This situation is depicted schematically in Figure 4.8.

According to Landau and Lifshitz (1970), the constitutive equation for the electrostatic interaction in dielectric materials, usually called the *Maxwell effect*,

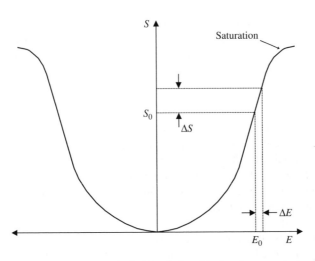

Figure 4.7   Strain versus electric field relationship in electrostrictive polymers.

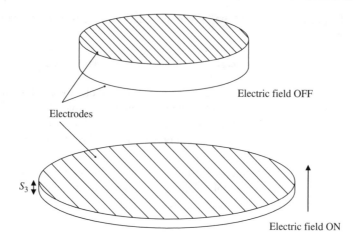

Figure 4.8   Schematic representation of dielectric elastomer actuation, electric field–induced contraction in the field direction and expansion in the perpendicular plane.

is, in tensor notation:

$$T_{ij} = -E_i D_j + \tfrac{1}{2}\delta_{ij} D_k E_k \tag{4.14}$$

where $\delta_{ij} = 1$ for $i = j$ and $\delta_{ij} = 0$ otherwise. The above expression for the mechanical stress exerted on the dielectric polymer can be simplified in compact notation when working out the strain–electric field relationship:

$$S_3 = -\tfrac{1}{2}\epsilon(s_{11} - 2s_{12})E_3^2 \tag{4.15}$$

Thus, according to Equation 4.15, the strain developed in a dielectric polymer is proportional to the squared electric field and will result in contraction of the thickness (note the minus sign). The transverse effect is formulated mathematically as in the next equation:

$$S_1 = \tfrac{1}{2}\epsilon s_{11} E_3^2 \tag{4.16}$$

For practical use of dielectric elastomers, the polymers employed must have a low elastic modulus. Silicones (polydimethyl siloxane) and acrylic polymers are generally used. The most commonly used are three commercially available materials (Madden (2004)), namely HS3 silicone (Dow Corning), CF 19–2186 silicone (Nusil) and VHB 4910 acrylic (3M). Typical elastic modulus values for dielectric elastomers vary between 1 MPa and 10 MPa, and the required electric field is of the order of ≈150 MV/m.

Since electrostrictive polymers are also dielectric materials, the electrostrictive effect is superimposed on the Maxwell effect in most cases. It has been reported (Sommer-Larsen and Kornbluh (2004)) that up to 50% of the strain developed in electrostrictive polymers can be due to the Maxwell effect. However, in most electrostrictive polymers, the high elastic modulus (sometimes as much as 100 times higher than in dielectric elastomers) limits the effect of Maxwell forces.

Unlike other EAP technologies, the basic material for a dielectric elastomer actuator can be selected according to the application requirements. This gives the designer the flexibility to adapt the design to the particular application.

Dielectric elastomers feature high strains, up to 50% in the thickness direction, and larger than 100% in the transverse direction. The efficiency of the actuators can be as high as 30% if appropriate means of energy recovery are provided, Madden (2004).

It has been shown that in theory these actuators offer the largest elastic energy density of all field-responsive actuators (Sommer-Larsen and Kornbluh (2004)). Practical values for these and other EAP actuators can be reduced by a factor of 10 to 100 when the accompanying components (for instance packaging) are considered in a particular application.

Dielectric elastomers can sustain a frequency bandwidth up to 1 kHz for some silicone materials. The electrical equivalent load for dielectric elastomers is capacitive and resistive, and, therefore, one of the limiting factors in the rate of response of these actuators is the charging time.

Of the various shortcomings of using dielectric elastomers, it is worth noting the high electric fields required for operation, which can lead to dielectric breakdown. Dielectric breakdown also imposes a practical limit on the maximum stress that the actuators can sustain, which is of the order of 1 MPa.

On the basis of the above-described features and limitations, dielectric elastomers have been proposed as suitable for use as actuators in robotics applications. Also, given the large stroke available with these actuators, they could potentially replace industrial linear actuator technologies such as pneumatic cylinders or electromagnetic solenoid actuators.

Another interesting aspect (see Sommer-Larsen and Kornbluh (2004)) is that because of their low elastic modulus, dielectric actuators are impedance matched with air. This opens up the potential field of application of dielectric elastomers to diaphragm pumps or loudspeakers.

4.2 Design issues

Design using EAP actuators is an open research area. Hitherto, design concepts have been imported from other emerging actuator technologies, but it is likely that specific concepts will have to be developed for EAP actuators. In this section, we briefly review open topics in designing with EAP technologies. The design of control strategies to suit the actuation characteristics of each technology is analyzed in the next section.

1. *Selecting the most appropriate actuator.* An application- oriented selection of a particular EAP technology (and if possible within a technology, a particular

polymer) is probably the best approach. The selection should not be limited to a matchup between actuator figures of merit and application requirements; it is recognized that other factors might play a very important role.

In aqueous environments (underwater applications or applications in body fluid environments), wet EAP actuators are probably the best option. In these environments, wet EAPs do not have to be hermetically packaged; contact with aqueous electrolytes is the ideal environment for them. Other factors, for instance biocompatibility, may determine the choice of a particular wet EAP actuator (for instance, CPs have proven to be biocompatible, unlike polymer gels and other wet technologies).

2. *Electrode design.* When EAP actuators have to be electroded (dielectric elastomer, electrostrictive, IPMC actuators) and are characterized by large actuation strain, defoliation is likely to result. Electrode design is still a field of study, and it is one of the critical aspects that have to be taken into consideration when designing high-stroke EAP actuators.

3. *Strong electric fields.* Electronic EAP actuators are characterized by fast response time and high stroke at the expense of strong driving electric fields. This feature is common to other emerging actuators (piezoelectric actuators and electrorheological fluid actuators in particular) and poses practical technological problems in connection with wiring and connector technologies. This is therefore a design issue as regards the application of dielectric elastomer and electrostrictive actuators.

4. *Packaging.* Ionic (wet) EAP actuators must be kept wet for operation; they stop functioning if they dry out. Hermetically packaging wet EAP technologies is a ticklish technological problem that must be overcome if wet EAP actuators are to be applied in dry environments.

5. *New design configuration and geometrical concepts.* Design concepts have been imported from other emerging actuator technologies. This is true for instance of CP actuators, which are configured in multilayered schemes (conceptually similar to multimorph piezoelectric actuators) for converting linear displacements into bending motion.

The opposite is true in the case of IPMCs, where bending motion is a natural product of direct actuation and where new configurations are required to convert bending motion into linear actuation.

4.3   Control of EAPs

This section analyzes EAPs with reference to control requirements. EAP actuators embrace dissimilar technologies that require different control approaches. Electronic EAP actuators (dielectric elastomers and electrostrictive polymer actuators) do not require complex control strategies.

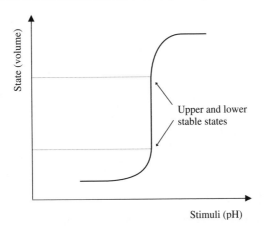

Figure 4.9 Discontinuous state transition in polymer gel actuators.

On the other hand, some wet EAP technologies exhibit complex nonlinear phenomena that impose severe demands on control strategies. This is true, for example, of IPMCs, whose dynamics are not the same upon contraction as they are upon relaxation (this is also true of all thermal actuators like SMAs). It is likewise true of some polymer gel actuators, which may exhibit discontinuous volume changes in response to slight alterations in the osmotic pressure (see Figure 4.9).

Chapter 1 introduced the biomimetic basis for switching control techniques in the context of muscle contraction control. In this case, switching techniques have proven to be good alternatives in dealing with the intrinsically changing dynamics of mammalian muscles on contraction and relaxation.

Power modulation in an actuator (or, alternatively, modulation of a state variable) can be achieved in some cases by means of switching techniques. By switching among discrete input levels, one can obtain averaged values of the actuator's output state. In switching techniques, the switching frequency and its value relative to the actuator's time constants are particularly important for optimum design.

Modulating the output state of an actuator by means of switching techniques produces a switching ripple around the reference trajectory (see Figure 4.10). If the switching frequency is higher than the bandwidth of an actuator, the average output over a switching period provides a good estimation of the true value of the output. Under these circumstances, a model of the averaged dynamics (in the case of an actuator with changing dynamics, i.e. IPMCs or SMAs) will suffice for the design of a controller to regulate the state variable by regulating the average (Mitwalli (1998)).

Let us consider the contraction and relaxation dynamics of an IPMC actuator that is mathematically described by two linear time-invariant models:

$$\dot{x}(t) = A_1 x(t) + B_1 u(t) \tag{4.17}$$

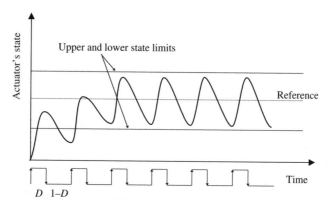

Figure 4.10   Effect of switching techniques in controlling varying-dynamics actuators: a ripple is produced around the reference; this is acceptable if it is lower than the upper and lower reference limits.

and,

$$\dot{x}(t) = A_2 x(t) + B_2 u(t) \tag{4.18}$$

where $x(t)$ is the state of the actuator and $u(t)$ is the input. A_i and B_i are constant matrices that describe the dynamics of the actuator upon contraction and relaxation.

Let us now assume two switching functions, $q(t)$ and $\bar{q}(t)$, of a fixed switching period, T. Let $\bar{q}(t)$ be the complement of $q(t)$ and let D be the duty cycle determining the time each switching function is active. The average of a function $y(t)$ over a switching period is defined as

$$\overline{y(t)} = \frac{1}{T} \int_{t-\frac{T}{2}}^{t+\frac{T}{2}} y(t)\, dt \tag{4.19}$$

The averaged model for the varying-dynamics actuator when the switching functions $q(t)$ and $\bar{q}(t)$ are applied to the input can be obtained by averaging Equations 4.17 and 4.18 and adding them together; note that the state matrices are constant:

$$\dot{\bar{x}}(t) = A_1 \overline{q(t)x(t)} + A_1 \overline{\bar{q}(t)x(t)} + B_1 \overline{q(t)u(t)} + B_1 \overline{\bar{q}(t)u(t)} \tag{4.20}$$

Assuming that the actuator's time constants are well above the time period of the switching signal, we may presume that the actuator states can be approximated by their averages over a switching period: $x(t) \approx \bar{x}(t)$ and $u(t) \approx \bar{u}(t)$. If this is true, then considering that $q(t) = D(t)$ and $\bar{q}(t) = 1 - D(t)$, it follows that

$$\dot{\bar{x}}(t) = \mathbf{A}\bar{x}(t) + \mathbf{B}\bar{u}(t) \tag{4.21}$$

where $\mathbf{A} = D(t)A_1 + (1 - D(t))A_2$ and $\mathbf{B} = D(t)B_1 + (1 - D(t))B_2$.

Equation 4.21 is the averaged model of the varying-dynamics switched actuator assuming the switching frequency is high enough as compared to the actuator's time constants.

In the case of IPMC actuators, the dynamics of contraction and relaxation are different, and, in general, switching frequencies can be made as fast as is desired since they have electrical input dynamics. This is not true of polymer gel actuators, where the input dynamics (for instance, acid–base mixing processes in pH- controlled drives) cannot generally be neglected.

A good switching control design for EAP actuators must consider the intrinsic dynamics of the actuator and the input system (Mitwalli (1998)).

4.4   Figures of merit of EAPs

EAP actuators are the newest of the actuation technologies covered in this book. Some of the technologies (see for instance CNTs) were discovered less than ten years ago. There are therefore no commercial or industrial applications of EAPs. EAP actuators are currently the subject of intense research.

One direct consequence of the novelty of these actuators is that there are no reliable experimental data comparable to the data available for other technologies. The data published for this emerging technology cannot readily be compared to those available for other technologies; indeed, in most cases, they are based on theoretical predictions. This is particularly true of the data on energy density or power density for EAP actuators: they are usually based on estimations of the maximum available elastic energy (by considering the material's Young's modulus and the available attainable strain).

The data presented here therefore represent a mere theoretical prediction, which will have to be verified experimentally in the coming years. It is a reasonable assumption that the experimental actuation characteristics of EAP actuators will be much lower than the theoretical predictions.

The heading of EAP actuators embraces technologies that are dissimilar, and as a result the actuation characteristics span a broad range. Ionic and electronic EAP actuators in particular differ basically in their operational frequency and power density. The following paragraphs analyze the static and dynamic performance characteristics of EAP technologies. These are compared with other emerging actuator technologies in Chapter 7.

4.4.1   Operational characteristics

Static performance

The maximum *relative stroke* exhibited by EAP actuators is larger than that of any other actuator technology. It may be as large as 300–400% in the case of dielectric elastomer actuators (both silicone and acrylic based) or as low as ≈5% in the case of CP actuators.

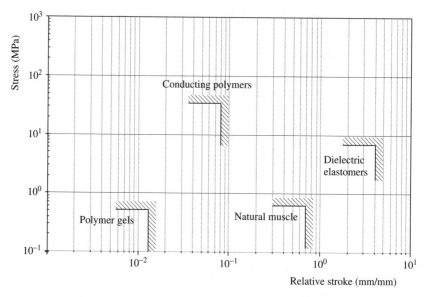

Figure 4.11   Maximum stress versus relative stroke for some EAP actuator technologies. Comparison with human muscle properties.

The *stress* attainable by EAP actuator technology is high in the case of CP actuators (of the order of 50 MPa) and low in all the other technologies. In the case of polymer gel actuators in particular, it can be as low as 1 MPa.

Figure 4.11 shows the relative position of some EAP technologies (based on theoretical expectations) in terms of maximum attainable stress and strain. Data for this comparison were gathered from scientific publications in the list of references at the end of this book.

Dynamic performance

EAP actuators are readily classified into slow and fast technologies. Ionic EAPs are slow technologies since their response time is diffusion-limited. CNT actuators are an exception to the above classification. Although they are ionic EAPs, CNT actuators are not diffusion-limited and hence respond faster than other wet EAP actuators.

The *work or energy density per cycle* of EAP actuators is high in the case of CNT actuators (≈ 50 J/cm^3) and is lower in all the other technologies. In the case of polymer gel actuators in particular, it is very low ($\leq 10^{-1}$ J/cm^3).

The response time is lower in electronic EAP actuators than in wet EAP technologies. Consequently, the frequency bandwidth is higher in dry EAP actuators. This means that values are high for the power density of CNT actuators ($\leq 10^3$ W/cm^3), electrostrictive polymer actuators ($\leq 10^4$ W/cm^3) and dielectric elastomer actuators ($\leq 10^3$ W/cm^3).

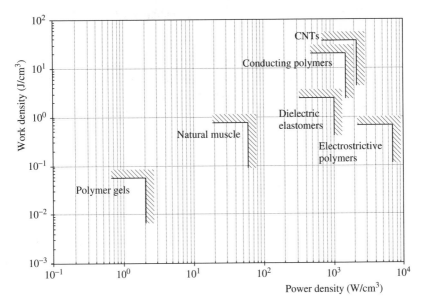

Figure 4.12   Maximum energy density per cycle versus power density for some EAP actuator technologies. Comparison with human muscle properties.

The relative position of EAP technologies in the dynamic plane (work density versus power density) is shown in Figure 4.12. The relative position of EAP actuators in the dynamic plane is compared with that of other emerging actuator technologies in Chapter 7.

4.4.2   Scaling laws for EAPs

The analysis of scaling laws for EAP technologies is similar to the analysis performed for other emerging actuator technologies. In both ionic and electronic EAP actuator technologies, the available force is a function of the cross-sectional area of the actuator. The area scales as the square of the dominant dimension, L, and thus $F \propto A \propto L^2$.

The stroke in EAP actuators is usually given as a percentage of the actuator's length. The maximum stroke of the actuator is thus proportional to the actuator's length: $S \propto L$.

The energy density per cycle can be derived as the product of force and stroke and is thus proportional to the cube of the dominant dimension: $W \propto L^3$. The energy density will remain constant upon scaling, $W_V = W/V \propto L^0$.

The main difference in the scaling properties of EAP actuators lies in the time response, frequency of operation and power density. In diffusion-limited actuators (all ionic EAP actuators except CNT actuators), the time response is given by

Table 4.2   Operational characteristics and scaling trends of EAP actuators.

Figures of merit	Wet EAP actuators	Dry EAP actuators
Stress, F	Up to 10^2 MPa for CPs	Up to 10 MPa for DEs
Strain, S	Up to 300–400% for dielectric elastomer actuators	
Work density, W_V	Up to 50 J/cm^3 for CNTs	
Power density, P_V	Up to 10^4 W/cm^3 for EPs	
Scaling trends		
Force	$F \propto L^2$	
Stroke	$S \propto L$	
Work per cycle	$W \propto L^3$	
Energy density	$W_V \propto L^0$	
Bandwidth	$f \propto L^{-2}$	$f \propto L^{-1}$
Power density	$P_V \propto L^{-2}$	$P_V \propto L^{-1}$ (and CNTs)

Equation 4.2:

$$t = \frac{x^2}{2D} \propto L^2 \qquad (4.22)$$

Equation 4.22, then, indicates that the time response of ionic EAPs scales down rapidly as the actuator's dimensions are reduced. Since time response and frequency bandwidth are inversely proportional, it follows that $f \propto L^{-2}$.

For electronic EAPs and CNT actuators, the time response is proportional to the actuator's dominant dimension, $t \propto L$. This means that the frequency bandwidth is inversely proportional to the actuator's dimensions, $f \propto L^{-1}$.

Since the maximum power density of an actuator is the energy density per cycle times the maximum operational frequency, it follows that the power density scales as the frequency, that is, $P_V \propto L^{-2}$ for wet EAP actuators and $P_V \propto L^{-1}$ for dry actuators.

Table 4.2 summarizes the scaling properties of EAP actuators.

4.5   Applications

Case Study 4.1:   Conducting polymer actuators as blood vessel connectors

We have seen in this chapter that CP actuators, also known as conjugated polymer actuators, have potential as an actuator technology in biomedical applications (see also Smela (2003)).

In body fluids (blood, urine), CP actuators find the standard environment they require for operation. CP actuators are ionic polymer actuators and as such require to be immersed in an electrolyte (for instance, blood or urine). Thus, where other actuator technologies need protection against such corrosive environments, CP actuators find an intrinsically suitable working environment.

Several research centers and firms are currently working on the application of CP actuators in biomedical environments. To mention only a few examples, Infinite Biomedical Technologies (2004) is in the process of developing a PPy-based microvalve for prevention of urinary incontinence and Intelligent Polymer Research Institute (2004) is working to improve the mechanical properties of cochlear implants and to allow sensory feedback to assist surgeons during implantation.

This case study explores the application of CP actuators in the process of surgically connecting severed small blood vessels, an operation known as microanastomosis. The process of reconnecting severed small blood vessels, commonly applied in brain surgery, is a delicate and challenging task since it involves lengthy operations and commonly leads to adverse reactions.

The approach is currently being developed in cooperation between the Mechanical Engineering Department at the University of Maryland and Micromuscle AB (http://www.micromuscle.com/), a company located in Sweden. The goodness of the approach is underpinned by the biocompatibility (both *in vitro* and *in vivo*) of CP actuators.

The new blood vessel connector is based on a bimorph configuration in which a microfabricated PPy bilayer is rolled up into a small-diameter cylinder by the permanent application of a reducing potential. See Figure 4.13 for a schematic representation of the rolling process in the bilayer actuator.

The dimensions of the electrically reduced bilayer actuator are smaller than the diameter of the severed blood vessel. During the surgical operation, the CP actuator is inserted between the two severed ends while being actuated, that is, its diameter is reduced. Once it is in place, the electrical potential is removed, the

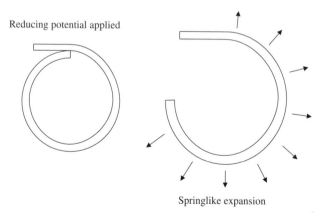

Reducing potential applied

Springlike expansion

Figure 4.13 A bilayer CP actuator bends upon application of a reducing potential. If it is not activated, the springlike expansion can be used to exert internal pressure on the severed vessel.

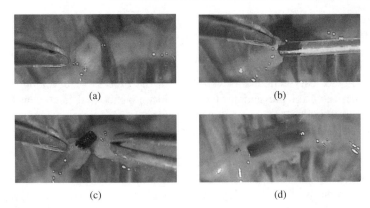

(a)   (b)

(c)   (d)

Figure 4.14   Steps in microanastomosis operations with CP actuators (Reproduced by permission of WILEY-VCH).

actuator is oxidized and it expands, clamping both ends of the severed vessel and keeping them together. See Figure 4.14 for the steps in surgical implantation of the CP connector.

This is a typical example of a one-time actuation mechanism. It is the springlike characteristics of the CP actuator in the oxidized state that keep the vessel in place. The CP actuator is incorporated into the vessel tissue. Because the bimorph CP actuator is very thin, it does not restrict blood flow. The complete surgical process requires slow actuation characteristics, which again match those of CP actuator technologies.

Case Study 4.2:   Tactile displays based on CP actuators

Tactile displays for visually impaired persons is an active area of research, particularly the part relating to actuation technology. Take for instance the case of active tactile braille lines to develop computer interfaces for the blind. Each braille character comprises six active dots (usually referred to as 'tactels' or 'taxels'). Each tactel must be independently activated in response to the cursor position on the screen so that the blind person can feel the text on the screen.

Typical tactel displacements are in the range of 0.2 to 0.5 mm. The six active dots are configured in a 3×2 matrix, with each dot positioned just a few millimeters from the neighboring dots. Refresh time for the braille line must be fast to allow quick reading.

There are commercial active braille lines, based on piezoelectric multimorph actuators. The actuation characteristics of these actuators meet the specifications noted above, but they make interfacing costly (a typical braille line requires 80 characters, which makes 480 independent actuators).

Research has focused on alternative actuation technologies; for instance, shape memory alloy actuators have been proposed as alternative solutions. Bistable SMA

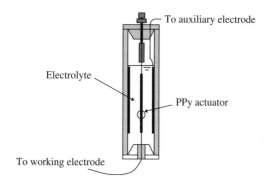

Figure 4.15   An actuator concept based on EAP CP actuators for tactile displays (Reproduced by permission of WILEY-VCH).

actuators (Brenner *et al.* (2000)) have been proposed in an attempt to reduce the inherently high energy consumption of thermal actuators. Fatigue, like energy consumption and efficiency, also imposes strict requirements on SMA actuators.

EAP actuators have also been proposed as an alternative actuation technology in the context of tactile displays (Wallace *et al.* (2002)). The concept is based on a hollow PPy tube, which is bias loaded with a linear spring (see Figure 4.15). A Platinum electrode is wound around the entire length of the PPy tube, helping to stiffen it and allowing electrical contact along the actuator.

The PPy tube is immersed in an electrolyte and the corresponding counterelectrodes are included in the hermetically sealed package. The configuration is such that when the PPy actuator is not activated, the bias spring keeps the tactel raised. Upon activation, the tactile dot is pulled down.

The overall dimensions of the actuator (approximately 70 mm in length and 10 mm wide) are not yet compatible with the application (see Figure 4.15). The application is cited here simply to illustrate the development of linear actuators based on EAP technologies.

Case Study 4.3:   CP-based microvalve design

Microfluidics is an active research domain with particular application in the context of biotechnologies. The concept of Lab-on-Chip has been coined to denote the technological approach that combines microfluidic technologies, biochemical sensors and processing to develop molecular or chemical laboratories at an integrated level.

A crucial component in this concept is the actuator, which performs valve functions and/or pump functions in such a device. In view of the high level of integration required, several technologies have been proposed, and in every case the scaling laws play a crucial role in the final selection of the most appropriate one.

EAP technologies, particularly wet EAP actuators, are characterized by large volume changes during actuation. In planar configurations, out-of-plane volume

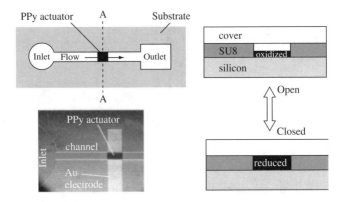

Figure 4.16   Operation of a PPy actuator in valve mode: the PPy film is deposited between the inlet and outlet in the fluid channel (top left) and electroded (bottom left). The volume change from the oxidized to the reduced state triggers the closing function (right) (Reproduced by permission of WILEY-VCH).

changes larger than 30% have been reported in the case of CP actuators (Smela and Gadegaard (2001)).

The out-of-plane volume change can be exploited to develop active valves for applications of this kind. Figure 4.16 shows a schematic representation of the valve concept and the actuation mechanism. PPy is deposited in the channel between the inlet and outlet. When the PPy actuator is in the oxidized state, its volume is kept to a minimum. Upon electrically induced reduction, the out-of-plane volume change is used to close the channel between inlet and outlet.

The concept can be extended to develop a peristaltic pump. This involves the combination of several valves as described above.

5

Magnetostrictive actuators (MSs)

Magnetostrictive actuators are based on the Joule effect (magnetostriction), the oldest of the transduction phenomena described in this book as a basis for emerging actuator technologies. Magnetostriction determines high-force, relatively low-stroke actuators with attainable frequency bandwidths up to several kilohertz. Given this type of performance, MS actuators are mostly suited to active vibration control applications. This chapter therefore addresses two parallel issues: magnetostrictive actuators as an emerging technology, and active vibration control and smart structures as the paradigmatic application of these actuators.

The chapter begins with an introductory historical note on the discovery of magnetostriction and a description of the basic magnetic properties of materials (fully applicable to MRF actuators as described in Chapter 6).

Magnetostriction is described and analyzed in detail in the subsequent sections. At the same time, issues relating to positive and negative magnetostriction, concomitant effects like the twofold change in the Young's modulus of these materials, and properties of magnetostrictive materials are introduced.

There is a section devoted specifically to the mechatronic design of MS actuators, particularly, design for improved stroke, linearized operation and selected dynamic properties of actuators. In the section devoted to electric and magnetic circuit design, the reader is referred to Chapter 6, where these aspects are dealt with in more detail in the context of MRF actuators.

The section on control of MS actuators offers a detailed analysis of the applicability of these actuators to active vibration control and smart structures. We decided also to include here those technologies that are most suitable for developing the concept of smart structures and smart actuators, for example, piezoelectric and

Emerging Actuator Technologies: A Micromechatronic Approach   J. L. Pons
© 2005 John Wiley & Sons, Ltd

shape memory alloy (SMA) actuators. As a result, smart structures are dealt with in a unified manner in this chapter rather than in separate sections in Chapters 2 and 3.

As in previous chapters, the various static and dynamic figures of merit for this technology are analyzed. Since MS actuators are potential competitors of piezo-electric stacks in several applications, a reference is included to the performance characteristics of piezoelectric stacks. The scaling trends for the various different figures of merit are introduced at this point.

Finally, this chapter introduces application cases for MS technologies. All of these are in the field of vibration control and smart structures, as these constitute the typical application field for this technology.

5.1 Principles of magnetostriction

5.1.1 Historical perspective

Magnetostriction is one form of energy transduction from the magnetic domain to the mechanical domain. It is a phenomenon observed in all ferromagnetic materials. The following sections analyze in detail the basics of the magnetic properties of materials and, in particular, of the magnetostrictive effect.

The magnetostrictive effect, also called *Joule effect*, was first described by J.P. Joule (1818–1889) in 1842 after the observation of a change in length of nickel upon the application of an external magnetic field. Following this discovery, nickel and its alloys were widely used during the 1940s and 1950s both in military and civil applications. The low magnetostrictive effect in nickel (typically of the order of 50 ppm), however, effectively limited the scope for new applications.

In later years, terbium (Tb) and dysprosium (Dy) were characterized as elements exhibiting a strong magnetostrictive effect at low temperatures, between 100 and 1000 times that of nickel. The addition of iron as an alloying element made the strong magnetostrictive properties of terbium and dysprosium available at ambient temperature. The critical factor and limiting parameter for the application of $TbFe_2$ and $DyFe_2$ proved to be the high-strength magnetic fields required to achieve a high strain.

At about the same time, researchers at the US *Naval Ordnance Laboratory*, today the *Naval Surface Warfare Center*, began to develop lanthanide alloys, lead-ing eventually to the discovery of Tb–Dy–Fe alloys with *giant* magnetostrictive properties (of the order of 1500–2000 ppm). The compound was called *Terfenol-D*: **TER** for Terbium, **FE** for Iron, **NOL** for Naval Ordnance Laboratory (NOL) and **D** for Dysprosium.

Magnetostriction, as the conversion of energy between the magnetic and the mechanical domain, is accompanied by several interesting phenomena. The reverse magnetostrictive effect is also known as the *Villari* effect and accounts for the conversion between mechanical and magnetic energy.

There is a particular instance of the magnetostrictive effect known as the *Wiedemann* effect. The Wiedemann effect is the conversion from a helical magnetic field to torsional mechanical energy. It typically occurs when a permanent axial magnetic field is superimposed on the magnetic field generated by an electrical current flowing axially in a ferromagnetic probe. The converse effect, that is, the generation of a helical magnetic field upon application of a torque, is known as the *inverse Wiedemann effect* or *Matteuci effect*.

Of the various different transduction processes in either direction between magnetic and mechanical energy domains, magnetostriction is the one typically employed in actuator development. It may be described as the analogous effect to piezoelectricity in the magnetic domain. However, there are a number of specific points on which they differ.

5.1.2 Basics of magnetic properties of materials

Usually, analysis of the magnetic state of materials is based on so-called *magnetization curves*. The magnetization curve for a particular material plots the *magnetic flux density, B* as a function of the *magnetic field strength, H*. The magnetic flux density, B, may be viewed as the magnetic counterpart of the electrical displacement or charge density, D. The magnetic field strength, H, and the electric field, E, are likewise analogous.

The magnetization curve for the vacuum represents a linear relationship between magnetic field strength and magnetic flux density (see Figure 5.1a). The slope of the magnetization curve for the vacuum is called the *permeability*, μ_0. Similarly, the permeability of a given material, μ, is the ratio of the internal magnetization of that particular material in response to an applied magnetic field.

The *relative permeability* of a material, μ_r, is defined as the ratio of the magnetic permeability to the vacuum permeability:

$$\mu_r = \frac{\mu}{\mu_0} \tag{5.1}$$

According to the shape of the magnetization curve, materials can be classified into *diamagnetic substances*, *paramagnetic materials* and *magnetizable materials*:

1. *Diamagnetic substances.* The small magnetic dipoles (due to electron spin) in diamagnetic materials cancel one another out. When an external magnetic field is applied, these small magnetic dipoles align themselves to oppose the external field. Consequently, the magnetization curve for diamagnetic materials is slightly below the magnetization curve in vacuum (see Figure 5.1b).

2. *Paramagnetic materials.* In these materials, the small electronic magnetic dipoles do not cancel out exactly and, thus, exhibit a small permanent magnetic moment. When an external magnetic field is applied, this permanent magnetization intensifies slightly, and the magnetization curve for a paramagnetic material is situated slightly above the magnetization curve for vacuum.

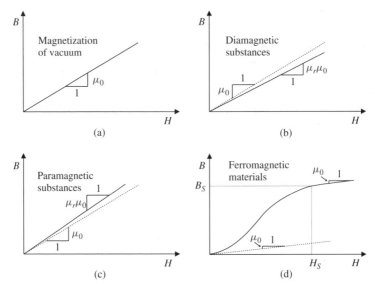

Figure 5.1 Magnetization curves for (a) vacuum, (b) diamagnetic substances ($\mu_r < 1$), (c) paramagnetic substances ($\mu_r > 1$), and (d) ferromagnetic materials, $\mu_r \gg 1$ (saturation at B_S).

3. *Magnetizable materials.* In these materials, the net magnetic moment can be made to point in a particular direction. The most relevant magnetizable materials for the purposes of actuators are *ferromagnetic materials.*

Ferromagnetic materials are iron, nickel, cobalt, manganese and their alloys. They can be permanently magnetized by the application of an external magnetic field. In ferromagnetic materials, the density of magnetic moments is known as *magnetization* or *magnetic polarization*, $\vec{M}$, and may be considered analogous to the electric polarization, $\vec{P}$.

Analogously to the electrical state of a dielectric material, the magnetic state of a ferromagnetic material can be described by means of two independent variables plus a third dependent one. If magnetic field strength, $\vec{H}$, and magnetization, $\vec{M}$, are the independent variables, the equation describing the magnetic estate of a ferromagnetic material will be

$$\vec{B} = \mu\vec{H} + \vec{M} \tag{5.2}$$

The process of magnetization in ferromagnetic materials involves the reorientation of magnetic domains so that they become aligned with the external magnetic field. Initially, most of the magnetic domains in a ferromagnetic material are oriented randomly. It is therefore relatively easy to orient them upon the application of an external magnetic field. Consequently, the slope of the magnetization curve is high for low values of the external field.

As the magnetic field increases, less magnetic domains are left for reorientation. As a result, it becomes increasingly more difficult to achieve a higher magnetic flux density. At this stage, the apparent permeability is reduced to the vacuum permeability. This situation is depicted in Figure 5.1d.

An interesting phenomenon of magnetism is the intrinsic hysteretic behavior of ferromagnetic materials in the B–H curve. Ferromagnetic materials tend to remain magnetized once the magnetic field is removed. The *magnetic remnance*, B_R, is defined as the remaining magnetization of a ferromagnetic material when the driving magnetic field is completely removed. The *coercivity*, H_C, of the ferromagnetic material is defined as the reverse magnetic field required to drive the magnetization to zero after having been saturated. Remnance and coercivity are depicted in Figure 5.2.

5.1.3   Magnetostriction: constitutive equations

Magnetostriction is the phenomenon whereby magnetic domains in a ferromagnetic material are reoriented and aligned in response to an applied external magnetic field. As a consequence of the magnetoelastic coupling in these materials, there is a macroscopic change in length in the direction of magnetization.

We described magnetostriction as being the analog of piezoelectricity in the magnetic domain. In fact, magnetostriction is a process of transduction between elastic mechanical energy (strain) and magnetic energy. However, there are evident significant differences between the two phenomena. The equations governing the magnetostrictive effect, *magnetostriction constitutive equations*, contain both linear

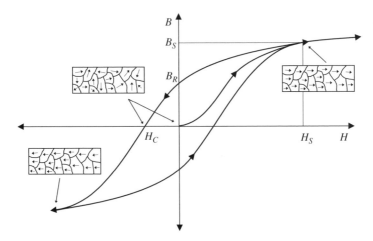

Figure 5.2   Hysteretic magnetization curve for a ferromagnetic material and relevant points: B_R, remnance; (H_S, B_S), saturation; and H_C, coercitive field strength.

and quadratic terms in the magnetic field strength. In tensor notation this is

$$S_{ij} = c^H_{ijkl}T_{kl} + d_{mij}H_m + m_{ijkl}H_kH_l$$

$$B_k = d_{kij}T_{ij} + \mu^T_{km}H_m \tag{5.3}$$

where S_{ij} is the mechanical strain, T_{kl} is the mechanical stress, c^H_{ijkl} is the mechanical compliance under zero magnetic field ($H = 0$), H_m is the magnetic field strength, μ^T_{km} is the magnetic permeability under constant mechanical stress, d_{mij} are the piezomagnetic displacement coefficients coupling linearly magnetic and mechanical variables, m_{ijkl} is the magnetostrictive coefficient coupling quadratically magnetic and mechanical variables and B_k is the magnetic flux density.

The equation coupling strain to magnetic field strength can be obtained from thermodynamic potential functions, and according to Equation 5.3 it has the following form:

$$S \propto c_1H + c_2H^2 \tag{5.4}$$

In Equation 5.4, c_1 defines the *piezomagnetic effect*. In order for a material to exhibit piezomagnetism, the crystal structure must be anisotropic. However, all ferromagnetic materials exhibit magnetostriction, that is, $c_2 \neq 0$. Therefore, the phenomenological description of piezomagnetism and magnetostriction is equivalent to the phenomenological description of piezoelectricity and electrostriction. The typical strain versus applied magnetic field curve for magnetostrictive materials is depicted in Figure 5.3. It shows the quadratic dependence of strain on magnetic field strength.

Carrying on with the analogy between electrical and magnetic domains, for a pure magnetostrictive material (exhibiting crystal symmetry and thus no piezomagnetism), the magnetomechanical coupling will be described by

$$S_{ij} = m_{ijkl}H_kH_l \tag{5.5}$$

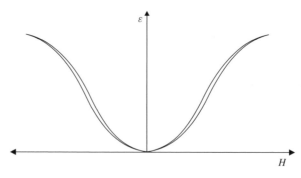

Figure 5.3   Quadratic functional relation between strain and magnetic field intensity in magnetostrictive materials.

As in the case of electrostrictive polymer actuators, Equation 5.5 can now be expanded in Taylor series around a DC bias magnetic field, H_0, and for small variations of the magnetic field around the DC bias, we have

$$S_i = S_{i0} + \frac{dS}{dH}\bigg|_{H=H_0}(H - H_0) + \frac{1}{2!}\frac{d^2 S}{dH^2}\bigg|_{H=H_0}(H - H_0)^2 + \cdots \qquad (5.6)$$

which, for the direction of polarization, reduces to

$$\Delta S \approx 2m H_0 \Delta H \qquad (5.7)$$

Notice that Equation 5.5 represents a quadratic dependence of strain on magnetic field strength. Once the magnetostrictive coefficients are defined for a particular material, Equation 5.5 indicates that the magnetostrictive material will contract or expand when either positive or negative magnetic fields are applied.

Consequently, Equation 5.7 describes a way of converting unidirectional displacements in a magnetostrictive domain into two-directional displacements by means of a bias magnetic field strength, H_0, a technique widely adopted in the design of magnetostrictive actuators.

The process is schematically depicted in Figure 5.4. Owing to the quadratic relationship between strain and magnetic field, the driving frequency presents nonlinearity, and the rate of the strain is twice the rate of the applied magnetic field (black lines in Figure 5.4). When a bias DC magnetic field, H_0, is applied, the strain becomes quasilinear around the bias magnetic field and the nonlinearity in the frequency is eliminated (grey, dashed lines in Figure 5.4).

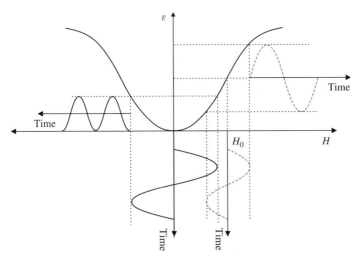

Figure 5.4   Linearization and two-directional operation with magnetostrictive materials.

5.2 Magnetostrictive materials: giant magnetostriction

Magnetostrictive materials expand or contract in the presence of a magnetic field because of the magnetostrictive effect described in the previous section. Of all magnetostrictive materials, the most useful for the implementation of actuators is Terfenol-D. Terfenol-D exhibits a magnetostrictive deformation of the order of 1500–2000 ppm, far above that of other magnetostrictive materials at ambient temperature, although it is lower than that of terbium or dysprosium when driven below approximately 180 K.

Magnetostrictive materials are small-stroke, large-force, solid-state actuators. As a direct consequence of the low strain, a mechanical impedance matching (mechanical amplification stages) is required in most applications.

Magnetostrictive materials are ferromagnetic substances with very high magnetic permeability. In these materials, the Curie temperature is the transition point at which the material becomes ferromagnetic. Below the Curie temperature, they present spontaneous magnetization.

5.2.1 Positive versus negative magnetostriction: effect of the load

Magnetostrictive materials exhibit mechanical displacements as described qualitatively by Equation 5.4, with $c_1 = 0$. In Equation 5.4, c_2 may take either positive, $c_2 > 0$, or negative values, $c_2 < 0$, producing positive or negative magnetostriction respectively. Nickel is an example of a material exhibiting negative magnetostriction. Iron will exhibit either positive or negative magnetostriction, depending on the crystallographic direction, Pérez-Aparicio and Sosa (2004).

Positive magnetostriction causes an increase in the material's length and a reduction in the transverse dimensions, so that the volume remains constant. On the other hand, negative magnetostriction causes a reduction in the material's length accompanied by an increase in transverse dimensions (see Figure 5.5).

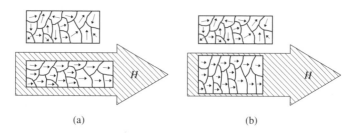

(a)   (b)

Figure 5.5   Change in strain due to magnetostriction: (a) positive magnetostriction increases length, while (b) negative magnetostriction results in contraction.

In magnetostrictive materials, the converse effect is also found. When a mechanical load is applied, there will be a change in the magnetization state of the material. This effect tends to decrease the reaction force against the external stress:

1. In negative magnetostrictive materials, a traction force will result in a reorientation of magnetic domains in a direction perpendicular to the applied force, that is, in the direction of the field that minimizes the reaction against the external force. In the presence of a compression force, the domains will tend to align in the direction of that force.

2. The opposite situation occurs in a positive magnetostrictive material. Compression forces will result in domains aligned in a direction perpendicular to the force, and traction forces will align the domains in their direction. This effect can be seen schematically in Figure 5.6.

The magnetoelastic coupling in magnetostrictive materials can be used to improve the stroke characteristics of MS actuators. This is illustrated qualitatively in Figure 5.7. For a positive magnetostrictive material, a compressive load causes the orientation of domains perpendicularly to the force. If such a material is mechanically preloaded, additional magnetic domains are left to be reoriented upon the application of an external magnetic field. This in turn results in an increased stroke.

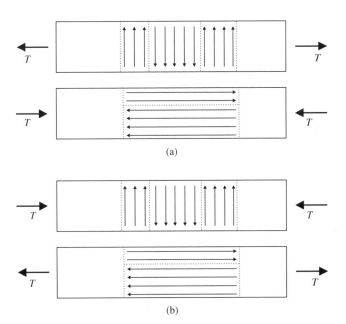

Figure 5.6 Effect of the load on magnetization: (a) negative magnetostrictive materials and (b) positive magnetostrictive materials.

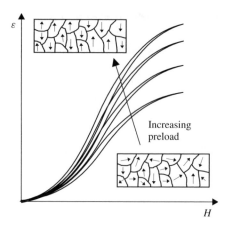

Figure 5.7   Stress-induced orientation of domains and subsequent increase in stroke.

5.2.2 ΔY-Effect in magnetostrictive materials

The Young's modulus, Y, in structural materials is a parameter used to define the material's stiffness. It is calculated as the ratio of change in stress to the corresponding change in strain in a given material.

In transducing materials, and in particular in magnetostrictive materials, there is a coupling between the stiffness (Young's modulus) and the magnetization state of the material. As a consequence, the Young's modulus for magnetostrictive materials is not constant but is rather a function of the magnetization state.

A change in the Young's modulus of a magnetostrictive material is commonly to be expected. Giant magnetostrictive materials undergo strains of the order of 1500–2000 ppm in response to changes in the magnetization state. The strains that these materials can exhibit when a pure mechanical load is applied are much lower that those produced by magnetostriction. Consequently, we may expect the effective elastic modulus, as the ratio of stress to strain, to be considerably affected by magnetization.

The so-called ΔY-effect is defined as the relative change in Young's modulus upon application of an external magnetic field, H, with respect to Young's modulus at zero magnetic field, $H = 0$ (Flatau *et al.* (1998)):

$$\Delta Y = \frac{Y_H - Y_0}{Y_0} \tag{5.8}$$

It is worth noting that Young's modulus undergoes a change even where the strength of the magnetic field is higher than the saturation magnetization. This indicates that the ΔY-effect cannot be convincingly explained only on the basis of the reorientation of magnetic domains due to external fields.

The maximum reported magnitude of the ΔY-effect (Clark and Savage (1975)) is of the order of 1, that is Young's modulus is doubled as a consequence of the

Table 5.1   Magnetostrictive and electrostrictive properties of some materials.

Property	Units	Terfenol-D	Hiperco	PZT-2
Mechanical properties				
Density	Kgm^{-3}	$9.25 \cdot 10^3$	$8.1 \cdot 10^3$	$7.5 \cdot 10^3$
Young's modulus, $H = 0$	GPa	26.5	206	110
Young's modulus, $B = 0$	GPa	55.0	-	60
Speed of sound	ms^{-1}	1690	4720	3100
Electrical properties				
Resistivity	10^{-6} Ω cm	60.0	0.23	0.01
Magnetostrictive and electrostrictive properties				
Permeability	–	9.3	75	1300
Curie temperature	C	387	1115	300
Maximum strain	ppm	1500–2000	40	400
Coupling factor	–	0.72	0.17	0.68
d_{33}	mA^{-1}, mV^{-1}	$1.7 \cdot 10^{-9}$	–	$300 \cdot 10^{-12}$
Energy density	Jcm^{-3}	14–$25 \cdot 10^{-3}$	–	10^3

change in the magnetization state of the magnetostrictive material. Moreover, since the resonance frequency of a magnetostrictive rod is

$$f_r \propto \sqrt{Y} \qquad (5.9)$$

it follows that the change in Young's modulus is related to the change in the resonant frequency squared. It will be seen in the coming sections that this property of magnetostrictive materials can be used in tunable vibration absorbers based on this technology.

Table 5.1 shows the properties of three magnetostrictive and electrostrictive materials. The table shows mechanical, thermal and electrical as well as electrostrictive and magnetostrictive properties.

5.3   Design of magnetostrictive actuators

The mechatronic design of magnetostrictive actuators typically addresses four main topics:

1. *Improvement of stroke.* Magnetostrictive actuators exhibit comparatively higher displacements than piezoelectric actuators. The displacement can be optimized by making use of the load effect on the magnetization state of the material.

2. *Linearized operation.* The constitutive equations of the magnetostrictive effect show a quadratic relationship between magnetic field and strain. Bias magnetic fields can be used to linearize this relationship.

3. *Push–pull operation.* As the strain versus magnetic field is a quadratic relationship, both positive and negative magnetic fields lead to unidirectional displacements. Bias magnetic fields can also be used to achieve two-directional
displacement.

4. *Optimization of electric and magnetic circuits.* The magnetostrictive actuator
is characterized electrically as an inductive load. The considerations are much
as in the case of magnetorheological fluid (MRF) actuators. And similarly
again, reluctance circuits must be optimized for efficient operation.

All these topics are addressed in the coming sections. In addition, the issue of
selecting the resonance frequency is analyzed as a typical procedure in developing
tunable vibration absorbers from magnetostrictive actuators. The typical design
configuration for magnetostrictive actuators is schematically depicted in Figure 5.8.
This includes prestressing mechanisms (usually based on compression springs), bias
magnetic field mechanisms (based on either permanent magnets or coils) and the
magnetic circuit.

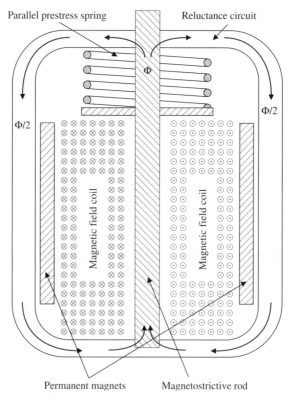

Figure 5.8   Schematic representation of the components in a magnetostrictive
actuator.

5.3.1 Design for improved stroke

The stroke of giant magnetostrictive materials is of the order of 1500–2000 ppm for static applications. These figures can reach up to 3000–4000 ppm when considering resonant amplifications, Claeyssen *et al.* (2002).

Terfenol-D rods generally consist of randomly oriented magnetic dipoles. The orientation of these dipoles upon the application of an external magnetic field produces magnetostrictive strain through magnetoelastic coupling. We noted earlier (see Section 5.2.1) that in positive magnetostrictive materials the application of compression loads causes magnetic dipoles to be oriented in the plane perpendicular to the load. This has the consequence of increasing the number of domains available for reorientation due to external magnetic fields, thus maximizing the stroke.

In practice, prestressing mechanisms are introduced. The prestress on the magnetostrictive rod can be applied by external springs either in parallel or in series, but the former is most commonly used.

A parallel prestressing mechanism is schematically depicted in Figure 5.8. Actuators with parallel prestressing are stiffer than actuators with series prestressing. Prestress values are typically of the order of 10–14 MPa.

Prestressing also has the beneficial effect of keeping the alloy working under compressive loads. If traction loads are avoided, the fatigue limit properties of magnetostrictive actuators are improved.

5.3.2 Design for linearized, push–pull operation

For the sake of convenience, let us recall Equation 5.4, which describes the quadratic relationship between strain and stress for a magnetostrictive material.

$$S \propto c_1 H + c_2 H^2$$

Linearized operation is a desirable characteristic in any actuator as it makes for simplified control. A linearization technique equivalent to the one described in this section was described with reference to electrostrictive polymer actuators. Indeed, an analogy was introduced between electrostriction and magnetostriction.

If a bias magnetic field, H_0, is applied to the magnetostrictive material, a quasilinear relationship will, in theory, be found between relative strain, ΔS, and relative magnetic field, $\Delta H = H - H_0$ (see Equation 5.7).

The bias magnetic field can be applied either by means of permanent magnets or through a coil-based constant magnetic field. In most common implementations, permanent magnets are used (see Figure 5.8). This makes for greater efficiency.

When designing magnetic biasing concepts for magnetostrictive actuators, the issue of nonsymmetrical operation has to be addressed. In the event that the bias magnetic field is not centered in the magnetization curve of the magnetostrictive material, operation may be nonsymmetrical. This is illustrated schematically in Figure 5.9. A bias magnetic field shifted to the saturation range of the magnetostrictive material will produce higher negative relative strain.

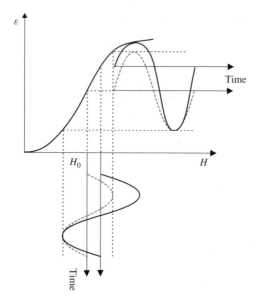

Figure 5.9   Effect of noncentered bias magnetic field on nonsymmetrical displacement.

5.3.3   Design of electric and magnetic circuits

Magnetostrictive actuators are characterized by exhibiting electrical inductive loads. This is mainly because coils are used to set up the magnetic field to drive the actuator. As such, the electrical driving requirements for MS actuators are equivalent to those of MRF actuators. We decided to concentrate on the analysis of electrical circuits for inductive loads in the chapter devoted to MRF actuators, and the reader is therefore referred to Section 6.2.3.

To establish the required magnetic field in the magnetostrictive material, the magnetic flux has to be guided to the active area. Again, this concept is implemented similarly in MRF actuators, and the basics of magnetic circuit design are analyzed in detail in Section 6.2.4.

The main difference between MS and MRF actuators as regards the magnetic circuit design lies in the use of permanent magnets in MS actuators to set up the bias magnetic field. The best configuration in terms of efficiency has been analyzed in detail by Janocha (2001).

Three elements are coaxially implemented to drive MS actuators: the MS material (T), the coil (C) and the permanent magnets (M). The specific configuration, in terms of the relative position of these three elements, has direct consequences for the homogeneity of the field in the MS material and for the attainable field strength. Optimum configurations are those in which the MS material (T) occupies the inner part of the actuator and the permanent magnets are placed in the outermost part of the actuators, that is, what is known as the **TCM** configuration (see Figure 5.8).

5.3.4 Design for selected resonance characteristics

A magnetostrictive actuator, as a second-order system, is characterized by a first resonance frequency at:

$$2\pi f_r = \sqrt{\frac{K_{eff}}{M_{eff}}} \tag{5.10}$$

where K_{eff} is the effective stiffness of the actuator and M_{eff} is the effective mass of the system. The effective stiffness of the actuator will, generally speaking, be a function of the geometry and material characteristics of the magnetostrictive rod, K_R, and of the prestress mechanism, K_P, being implemented.

If a cylindrical configuration for the magnetostrictive rod is assumed, its stiffness can be expressed as

$$K_R = \frac{Y A_{eff}}{L_{eff}} \tag{5.11}$$

In Equation 5.11, Y is the Young's modulus of the material and A_{eff} and L_{eff} are the effective cross-sectional area and length of the rod respectively.

The effective stiffness of the actuator, K_{eff}, will be greater than the stiffness of the magnetostrictive rod if the prestressing mechanism is placed in parallel to the magnetostrictive material, $K_{eff} = K_R + K_P$. On the other hand, where the prestress mechanism is in series, the effective stiffness of the actuator is less than that of the magnetostrictive rod, $1/K_{eff} = 1/K_R + 1/K_P$.

The combination of the magnetostrictive material's Young's modulus (typically lower than that of piezoelectric materials) and the prestressing mechanisms leads to lower frequencies in magnetostrictive than in piezoelectric resonators. This was one of the distinctive characteristics that prompted the adoption of the magnetostrictive technology in Sonar applications.

The resonance frequency of the system can be adapted to the application by careful selection of the rod's geometry and the prestressing mechanism. If a dynamic adaptation is required (as in the application of MS actuators in tunable vibration absorbers), the ΔY-effect must be used.

As noted earlier, the ΔY-effect produces up to 100% variation in the Young's modulus of magnetostrictive materials. This in turn results in changes of up to 40% in the resonance frequency so that tuning can be effected in response to structural modal changes in vibration absorption applications.

5.4 Control of magnetostrictive actuators: vibration absorption

If there is one actuator technology that is almost exclusively linked to a single application, that is the magnetostrictive actuator, the application is active structural vibration control. Almost all the applications described in the literature on magnetostrictive actuators are related in one way or another to vibration suppression mechanisms.

Magnetostrictive actuators deliver high-output forces and relatively high displacements (compared to other emerging actuator technologies) and can be driven at high frequencies. These characteristics make them suitable for a variety of vibration control applications.

Other actuator technologies described in this book have been proposed in vibration control applications, in particular, piezoelectric actuators and shape memory actuators for active control and electro- and magnetorheological actuators in semiactive control. Semiactive vibration control is analyzed in detail in the context of ER and MR actuators (see Chapter 6.)

In this section, we address the topic of active vibration control as the paradigmatic application of magnetostrictive actuators. A thorough analysis of the topic would require a complete monograph, and so the discussion here is confined to a descriptive level. For further reading and analysis, the reader is thus referred to one of the various books available on the topic.

Most of the strategies discussed here are applicable to other actuator technologies, and there is frequent cross-referencing in the following paragraphs. A whole section is devoted to the topic of smart structures as a concept intrinsically linked to smart actuator technologies, and in particular to SMA, piezoelectric and MS actuators.

5.4.1 Active vibration suppression

Passive tuned vibration absorption is a classical approach to reducing undesired vibration in a structure. It consists in the use of a mechanical spring-mass-damper oscillator whose resonance frequency is tuned to the frequency at which structural vibration needs to be reduced. It has been successfully applied since the late-1800s (Flatau *et al.* (1998)) in the context of narrowband attenuation of undesired oscillations (see Figure 5.10a).

The development of actuator technologies has opened up the possibility of structural active vibration control. However, traditional actuator technologies impose various limits on the applicability of active vibration control:

1. *Limited actuator dynamics.* Limited dynamics means that vibration control is only possible in narrowband.

2. *Difficult integration.* Traditional actuators impose lumped systems as they cannot readily be integrated in the structure.

3. *Need for external sensors.* Traditional actuators are not susceptible of incorporating sensor functions, and so the smart actuator concept is hard to apply.

The advent of new actuator technologies extends the domain of vibration control to broadband and to responsive structures with integrated (smart) actuators. Returning to the example of the tuned vibration absorber in Figure 5.10a, it is

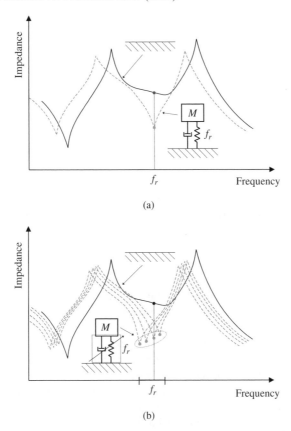

(a)

(b)

Figure 5.10   Vibration absorption: (a) tuned vibration absorber and (b) magneto-strictive tunable vibration absorber.

possible to implement the concept of a 'tunable' vibration absorber based on mag-netostrictive actuators. In this new concept, the ΔY-effect can be exploited to develop a vibration absorber whose resonant frequency can be tuned in response to structural modal variations (see Figure 5.10b and a more detailed description on page 190).

Active vibration control can be approached in *feedback* and *feed-forward* con-trol strategies. Feedback systems are further classified into *active damping control* and *modal-based feedback control*. The following paragraphs briefly describe the three approaches and outline the main characteristics.

Feed-forward vibration control

Feed-forward vibration control strategies can only be implemented in the case of known disturbances, or in the event of having a signal correlated to the disturbance.

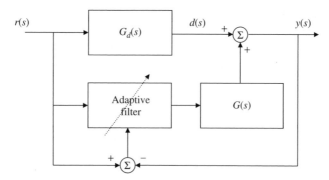

Figure 5.11   Adaptive feed-forward vibration control: the error between reference and output drives an adaptive filter to generate a second disturbance to cancel the primary one.

The underlying idea is to generate a second disturbance such that when applied to the plant, it cancels the effect of the primary disturbance.

In this approach, the reference signal (correlated to the disturbance) is filtered and applied to the plant. The error signal supplied by a sensor is used to adaptively tune the filter parameters so that the error is minimized. A block diagram of this vibration control approach is shown in Figure 5.11.

The feed-forward approach only guarantees vibration reduction at the sensor's location and so must be considered a local method. When sensor and actuator are not collocated, an estimate of the transfer function between sensor and actuator is required to compute the error signal that drives the filter parameter adaptation process. Since this process is based on an adaptive algorithm, it is robust and allows higher frequency band attenuation than feedback approaches.

Modal-based feedback vibration control

A general feedback control loop might be represented by the block diagram in Figure 5.12. In Figure 5.12, $G(s)$ is the plant whose vibration level must be kept bounded, $H(s)$ is a feedback compensator, $r(s)$ is a reference value of any plant variable (position, velocity) and $d(s)$ is an undesired disturbance that will affect the output $y(s)$.

The transfer function, $F_r(s)$, between reference input, $r(s)$, and plant output, $y(s)$, is readily available:

$$F_r(s) = \frac{y(s)}{r(s)} = \frac{H(s)G(s)}{1 + H(s)G(s)} \qquad (5.12)$$

The transfer function, $F_d(s)$, between the disturbance $d(s)$ and the system's output can be expressed as

$$F_d(s) = \frac{y(s)}{d(s)} = \frac{1}{1 + H(s)G(s)} \qquad (5.13)$$

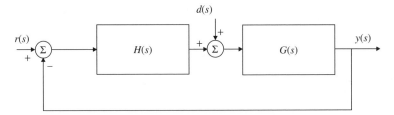

Figure 5.12   Feedback control loop.

It is evident that, when $H(s)G(s) \gg 1$, the output perfectly tracks the reference, that is, $F_r(s) \rightarrow 1$, and the effect of the disturbance on the output is negligible:

$$H(s)G(s) \gg 1 \quad \Rightarrow \quad y(s) \approx r(s) \quad \text{and} \quad \frac{y(s)}{d(s)} \approx 0 \qquad (5.14)$$

The modal-based feedback vibration control approach is intended to provide perfect tracking of the plant's reference value while ensuring high disturbance rejection.

According to Preumont (1997), irrespective of the method used to find the appropriate compensator, $H(s)$, the following characteristics are common to all modal reference feedback control systems:

1. They always depend on accurate models of the plant being controlled, and the bandwidth of the control system depends on the accuracy of the model. Low accuracy produces *spillover*, which is a reduced damping of residual modes.

2. Rejection of the disturbance in the control bandwidth leads to increased disturbance outside the control range.

3. A very high sampling frequency is required in digitally implemented control loops (of the order of 10 times the maximum control frequency).

Active damping

The objective of active damping control is to increase the damping of some of the vibration modes of a structure while leaving the system's transfer function unaffected outside the resonance modes. Ideally, the closed-loop transfer function $F(s)$ in Equation 5.12 would look the same as $G(s)$ out of resonance but would be more damped (the amplitude reduced) close to the resonance frequency.

According to Preumont (1997), active damping control can be achieved without a model of the structure and provide guaranteed stability under two conditions:

1. *Sensors and actuators are collocated*. In practice, this means that the sensor directly provides a measure of the action of the actuator. This is an important feature when considering smart structures (see Section 5.4.2).

2. *Actuator and sensor have perfect dynamics.* Perfect dynamics for sensor and actuator will produce infinite bandwidth rejection. The active damping system will have finite bandwidth to the extent that sensor and actuator have limited dynamics, which is always the case.

The condition of collocated sensors and actuators in an undamped structure produces alternating poles and zeros on the imaginary axis of the s-plane. If the system damping is moderately increased, the alternating pattern of poles and zeros shifts slightly toward the negative real half plane, guaranteeing stability.

Several active damping strategies can be envisaged (Preumont (1997)): velocity feedback, acceleration feedback, lead compensator position feedback and force feedback. Here, we briefly analyze active damping through velocity feedback, since this is of particular interest for active damping control implementations in smart structures.

Let us consider the feedback loop in Figure 5.12. In order to simplify the derivation, we will consider an undamped structure with the following governing equation:

$$\left[Ms^2 + K\right]x(s) = f(s) + Bu(s) \tag{5.15}$$

Since we are considering collocated sensors and actuators, the following equation applies:

$$sy(s) = B^T x(s) \tag{5.16}$$

Moreover, in the case of a velocity feedback law, the control action will be

$$u(s) = -Dsy(s) \tag{5.17}$$

If all three Equations 5.15–5.17 are combined, this gives

$$\left[Ms^2 + BDB^T s + K\right]x(s) = f(s) \tag{5.18}$$

The control action on the plant results in a viscously damped closed-loop transfer function. As discussed in the next section, in applications of piezoelectric actuators in smart structures, direct electrical current feedback can be considered equivalent in approach to direct velocity feedback and produces active structural damping.

'Tunable' vibration absorbers

Tuned vibration absorbers were mentioned earlier as a successful classic approach to vibration attenuation. As noted then, a spring-mass-damper mechanism is tuned so that its resonance frequency matches the particular frequency for which vibration reduction in a structure is desired.

The tuned vibration absorber is mounted on the structure whose vibration is to be attenuated. When the structure is excited at this particular frequency, the vibration is absorbed by the tuned mechanical system and the structure exhibits antiresonance (zero amplitude vibration) behavior (see Figure 5.10a).

Tuned vibration absorbers are limited to single-frequency vibration attenuation. If multifrequency attenuation is required, additional absorbers must be used. Magnetostrictive materials exhibit the ΔY-effect (see Section 5.2.2). This phenomenon produces changes in Young's modulus as the magnetization state of the material is modified. Since the resonance frequency of a second-order system is proportional to the square root of the Young's modulus, $f_r \propto \sqrt{Y}$, the ΔY-effect can be used to effectively change the resonance frequency of a magnetostrictive material.

This in turn can be used to implement a "tunable" vibration absorber. This is the way that an actuator with a seismic mass attached to it is developed (see the schematic representation in Figure 5.10b). By changing the magnetization state when an external magnetic field is applied to the magnetostrictive material, the resonance frequency of the tunable mass-spring-damper magnetostrictive system will be modified and can be adapted to requirements.

5.4.2   Smart actuators and smart structures

The most distinctive feature of smart actuators is their intrinsic ability to perform simultaneously as sensors and actuators. As was discussed in Chapter 1, this property is a product of the transducing characteristics of various materials.

The distinctive feature of smart structures is the high degree to which sensors and actuators are integrated as constitutive components of the structure. The integration of sensor and actuator functions results in functional and active structures.

Owing to the intimate combination of sensor, actuator and structural functions, they can perform the following roles:

1. *Structural health monitoring.* Embedded smart actuators can be used to assess the condition of a structure, for instance, to detect increasing loss of stiffness because of growing cracks.

2. *Failure prevention.* Upon detection of increasing loss of stiffness or in the event of overloading, smart actuators can be used to temporarily stiffen the structure.

3. *Integrity protection.* Smart actuators can generally provide reactive action following any undesired circumstance, for instance, when excessive vibration is sensed.

Of the various actuator technologies discussed in this book, SMA actuators, piezoelectric actuators and magnetostrictive actuators are the three best suited to develop the smart actuator and smart structure concepts. Their properties in this regard are briefly reviewed in the following paragraphs.

Shape memory alloy actuators in smart structures

SMA actuators can both sense the status (position) of and deliver mechanical energy to the structure in which they are embedded. They can be readily embedded in structures to provide functionality. Three possible actuation mechanisms are envisaged:

1. *Stiffening through control of the Young's modulus.* The Young's modulus of SMAs increases significantly upon transformation from martensite to the parent phase. This can be used to stiffen the structure transitorily. The stiffening process can be used, for instance, to shift the resonance frequency of the structure and avoid possible resonance vibration in response to external excitation.

2. *Generation of strain.* The shape memory effect can be used to exert a stress on the structure and strain it. This can also be readily used to change the resonance properties of the structure.

3. *Energy dissipation through the hysteretic pseudoelastic effect.* The pseudoelastic effect in SMAs (see Section 3.1.2) is characterized by a hysteretic dissipation phenomenon. Pseudoelasticity can be used to increase the apparent damping of the structure.

The first two actuation mechanisms can be implemented in response to some monitoring action that may also be accomplished by the SMA, this time performing as a sensor (see Section 3.3.3). The third mechanism is intrinsically passive.

Piezoelectric actuators in smart structures

Piezoelectric actuators were extensively discussed in Chapter 1. Like SMA actuators, they can perform both as sensors and as actuators. Unlike SMA actuators, piezoelectric actuators cannot be used to concomitantly impose and sense the same output variable (in Chapter 1, a method based on unbalanced bridge circuits was introduced as a possible approach to estimate this output variable), but they can sense the conjugate variable to the imposed one.

When discussing piezoelectric actuators, we pointed out (see Section 2.5.2) that current drawn is, in principle, proportional to the actuator's velocity. For the sake of convenience, let us recall these relationships here:

$$\Delta l \approx \Delta Q$$

$$v_p \approx i$$

$$a_p \approx \mathrm{d}i/\mathrm{d}t$$

According to the above equations, if control is achieved by feeding the current drawn directly back in a negative control loop, the result is equivalent to a direct velocity feedback approach. Direct velocity feedback, in a collocated sensor/actuator pair, leads to stable active damping control schemes (see section 5.4.1). Therefore, since a piezoelectric smart actuator is intrinsically collocated, this approach greatly facilitates the implementation of active vibration control in smart structures.

Also in Section 2.5.2, we discussed how the electrical boundary conditions applied to a piezoelectric actuator can result in *programmable actuator stiffness*. Programmable stiffness, like the first actuation mechanism in SMA actuators, can be applied for voluntary stiffening in response to monitoring functions.

Two actuation approaches can be envisaged for piezoelectric actuators:

1. *As linear actuators.* In this approach, they replace conventional actuators in controlling structure parameters. As noted earlier, this actuation approach is not suitable for development of the smart structure concept as sensor and actuator functions are lumped rather than distributed. We need discuss this scheme no further.

2. *As surface bonded actuators.* The piezoelectric material is laminated and bonded in thin layers to the structure (see Figure 5.13). This can serve for the application of either axial or bending loads.

Both actuation approaches are most commonly used to enhance the damping characteristics of the structure to which they are attached (first case) or embedded (second approach). As regards surface-bonded actuators, in controlling damping, again there are two possible strategies:

1. *Passive damping.* In this approach, the actuator is bonded to the structure and is electrically connected to a passive RLC circuit. The actuator acts as a voltage source when driven by the structural vibrations through the direct piezoelectric effect. The electrical energy is dissipated at the resistor. This results in an apparent increase of structural damping.

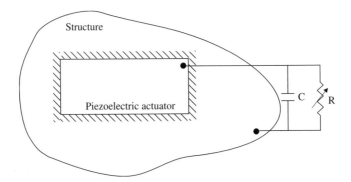

Figure 5.13 Piezoelectric actuator bonded to a structure and connected to an RLC circuit for passive damping.

Some selectivity in the damping process can be achieved by tuning the RLC characteristics of the passive circuit to the target frequency (see Figure 5.13).

2. *Active damping*. In this mode, the piezoelectric actuators are integrated in a modal reference feedback control strategy, in an active control strategy or in a feed-forward control strategy. In any of these control algorithms, the piezoelectric actuator can perform as a smart actuator, that is, sensing and actuating concomitantly.

Magnetostrictive actuators in smart structures

Magnetostrictive actuators are best suited for active vibration control of structures. Moreover, they are a typical example of a smart actuator. However, they are not suitable for integration in any type of structure.

In particular, planar configurations (similar to bonded laminate piezoelectric actuators (see Figure 5.13) are not ideal for magnetostrictive actuators. This is due to the difficulty in creating a uniform magnetic field for such a planar configuration.

Magnetostrictive actuators are more suitable for integration in discrete structures like three-dimensional trusses and structures with structural cables (bridges or buildings). There, they can take the place of passive structural elements and thus offer the possibility of cancelling out structural vibrations (see Bartlett *et al.* (2001)).

The issue of active control of truss structures has been comprehensively studied by Preumont (1997). We will not go into more detail here; for more details, the reader is referred to this or any other book on the subject.

Magnetostrictive actuators can be used in the context of vibration suppression in smart structures, in three different modes:

1. *Stiffeners*

2. *Dampers*

3. *Active elements.*

Stiffeners make use of the ΔY-effect, which allows a twofold increase in the actuator's stiffness. By changes induced in the structural stiffness of the structure in which they are integrated, the resonance frequency can be modified to prevent resonance amplification following external excitation.

Magnetostrictive actuators can also be used as dampers in active damping control approaches. Active damping control was already introduced in previous sections. The role of the actuator in this control scheme is to enhance the structural damping by means of direct velocity feedback or similar approaches.

Finally, they can be applied in active vibration cancellation control schemes, both feed-forward and feedback. In this approach, the MS actuator will provide the necessary secondary disturbance to cancel out vibrations. As in the case of piezoelectric or shape memory actuators, both sensing and driving functions can be implemented concomitantly. This is discussed in more detail in the next section.

5.4.3   Combined sensing and actuation

We described the reversibility of the magnetostrictive effect in previous sections. Two different phenomena, the Villari effect and the Matteuci effect, have been defined as transduction between force and torque (energy in the mechanical domain) to the magnetic domain.

Owing to the existence of these converse transduction processes, magnetostrictive transducers can be configured as sensors. Sensors based on either of the two above effects have been proposed to monitor various different physical variables. They are classified in three categories (see Dapino *et al.* (2002)):

1. *Passive sensors.* These rely on the transduction process itself, that is, the change in the material's magnetic properties in response to environmental changes.

2. *Active sensors.* These sensors use the magnetostrictive material in an active way. They are excited in a known manner, and this facilitates measurement of some properties of the material that change with an external physical variable.

3. *Combined sensors.* Here, the magnetostrictive transducer is used in combination with other materials. The magnetostrictive material is used to excite or alter another material which will, in turn, allow the target property to be measured.

In the context of concomitant sensing and actuation based on magnetostrictive materials, we are mainly interested in passive magnetostrictive sensors, that is, in the concomitant use of direct and converse effects for simultaneous sensing and actuation.

Let us recall here the linearized version of the constitutive equations (Equation 5.3) for the magnetostrictive effect:

$$S = c^H T + d H$$

$$B = \bar{d} T + \mu^T H \tag{5.19}$$

Equation 5.19 describes the coupling between magnetic and mechanical variables in the direct and converse magnetostrictive effect. The first equation describes the transducer as an actuator, that is, the resulting displacement is a function of the applied magnetic field strength. It further includes the coupling between mechanical variables, that is, the displacement resulting from mechanical load.

The second part of the constitutive equation describes the transducer in the role of sensor, relating the mechanical load to the magnetic induction. Again, this part includes the coupling between the magnetic field variables, that is, the applied magnetic field strength results in magnetic induction.

The two equations can be combined by solving the first part for H and substituting it in the second part to yield

$$B = T \left[\bar{d} + \frac{\mu^T c^H}{d} \right] + \frac{\mu^T}{d} S \tag{5.20}$$

Equation 5.20 describes the sensor model for the magnetostrictive transducer. It formulates the relationship between the resulting magnetic induction in the magnetostrictive material, B, the applied force, T, and displacement, S. Faraday's law can be used to determine the magnetic induction in the material. This states that the voltage induced in a coil wrapped around the magnetostrictive material is

$$V = NA \frac{dB}{dt} = NA \frac{dT}{dt} \left[\bar{d} + \frac{\mu^T c^H}{d} \right] + NA \frac{dS}{dt} \frac{\mu^T}{d} \tag{5.21}$$

where N is the number of turns in the coil and A is the cross-sectional area.

Equation 5.21 indicates that the voltage in such a sensor configuration is proportional to the rate of change of force (jerk) and to the rate of change of displacements (velocity) in the magnetostrictive material. It can be demonstrated (Dapino *et al.* (2002)) that the magnetostrictive process is fully reversible and $\bar{d} = d$, and that the term in Equation 5.21 corresponding to the velocity is one order of magnitude smaller than the terms relating to force. Therefore, for harmonic excitations, a model for the force measured by the transducer would be

$$T = \frac{V}{NA\omega} \left[\frac{1}{d + \frac{\mu^T c^H}{d}} \right] \tag{5.22}$$

The design of magnetostrictive transducers as sensors is equivalent to their design as actuators. Moreover, as we assumed that $\bar{d} = d$, this implies that if the transducer is designed for efficient operation as an actuator, it will be a high-sensitivity sensor. These are commonly used in sonar transducers where a transducer designed as an efficient emitter also yields the best results as a receiver.

For combined sensor and actuation operation of magnetostrictive actuators, the *bridge circuit configuration* as discussed in Section 1.5 can be implemented. To do this, the linearized constitutive equations for the magnetostrictive effect, Equation 5.19, must be rewritten in line with the electric-circuit analogy as discussed in Chapter 1.

We commence the process by multiplying the sensing part of Equation 5.19 by the actuator's cross-sectional area A. Considering that $\Phi = B \cdot A$, it follows that

$$\Phi = \bar{d} T A + \mu^T H A \tag{5.23}$$

Now, if we take the time derivative of Equation 5.23, multiply the equation by N (the number of turns in the MS actuator coil) and note that $V(t) = N d\Phi/dt$ and $H = I(t)N$, we obtain

$$V(t) = N\bar{d} \frac{dT}{dt} A + \mu^T N^2 \frac{A}{l} \frac{dI}{dt} \tag{5.24}$$

The Laplace transform of Equation 5.24 can now be developed. In addition, if we take into account low-frequency excitation of the transducer, we can assume that the material's deformation obeys Hooke's law ($T = Y\varepsilon$). Given these assumptions, we can write

$$V(j\omega) = N\bar{d}\frac{AY}{l}U(j\omega) + j\omega LI(j\omega) \qquad (5.25)$$

where Y is the Young's modulus and l is the length of the magnetostrictive actuator.

A simple inspection of Equations 5.25 and 1.5 will show that $Z_e = j\omega L$, that is, the blocked impedance of the magnetostrictive actuator (which is required to complete the second branch of the bridge circuit) is simply the inductance of the solenoid (with the MS material as the core) used to drive the MS actuator (see Figure 5.14).

A similar approach can be used to derive the second part of Equation 1.5. It can be shown (Pratt (1993)) that this produces

$$F(j\omega) = \frac{AY}{l}\frac{1}{j\omega}U(j\omega) + AY\, dNI(j\omega) \qquad (5.26)$$

If the solenoid inductance is used in the second branch of the bridge circuit, the voltage across the bridge ought to be proportional to the actuator's velocity. However, this is not the case, and the reason is that the blocked impedance is nonlinearly dependent on the current drawn. Consequently, concomitant sensing and actuation with MS actuators is difficult to achieve, although some authors (Pratt (1993)) have reported positive results in narrow frequency bands around the actuator's mechanical resonance.

5.5 Figures of merit of MS actuators

This section analyzes the performance characteristics of magnetostrictive actuators. Chapter 7 is devoted entirely to a comparative analysis of all emerging actuator

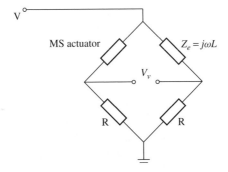

Figure 5.14   Bridge circuit configuration for concomitantly using an MS actuator as a sensor.

technologies, but here we pay particular attention to a comparative analysis of MS and piezoelectric actuators because these have the closest figures of merit. In this section, we also give some indications about scaling trends in MS actuators.

5.5.1 Operational range

Static performance

Magnetostrictive and piezoelectric stack actuators can be classified together as high-force devices. The absolute value of force in these actuators is in the kilonewton range ($F \geq 1$ kN).

Relative forces (for instance as compared to cross-sectional area or size), nevertheless, become several orders of magnitude lower than relative forces in piezoelectric actuators. This is mainly due to the accompanying components required to set up bias magnetic fields or to prestress the magnetostrictive actuator. While piezoelectric stack actuators can be directly applied to drive the load, MS actuators require coils to set up the magnetic field and, in most implementations, permanent magnets to apply bias fields.

Stroke in MS actuators is of the order of 1500–2000 ppm for static applications and close to 4000 ppm where resonance amplification is used in dynamic applications. In absolute value, the displacement is limited in practice to some tenths of a millimeter. Stroke is higher in MS actuators than in piezoelectric stack actuators, but the performance of piezoelectric stack and MS actuators is very similar in terms of the relative stroke (for instance, with respect to the length of the actuator).

Dynamic performance

The *energy density of magnetostrictive actuators* is in the range of $W_V \leq 10^{-3}$ J/cm^3 and must therefore be considered low. If only the magnetostrictive material were considered, the energy density would be closer to that of piezoelectric actuators, but, here again, the bulky accompanying elements cause reduced work density.

The *bandwidth of MS actuators* is high, $f \geq 1$ kHz. Together with piezoelectric stack actuators and some piezoelectric multimorph drives, they have the highest frequency bandwidth of all emerging actuator technologies. MS drives are driven at lower frequencies than piezoelectric stack actuators. There are two main reasons for this:

1. *Lower Young's modulus.* MS materials exhibit a lower Young's modulus than piezoelectric materials (two to three times lower). This results in lower resonance frequency of MS materials, which in turn limits the maximum driving frequency for actuators based on this technology.

2. *Eddy currents.* Changing magnetic fields induce electrical currents in the magnetostrictive materials; the higher the excitation frequency, the stronger

are the currents induced. This imposes a practical limit on the maximum frequency attainable with MS actuators.

Power density in MS actuators is a result of the two previous figures of merit. Since both energy density and bandwidth are lower in MS actuators than in piezo-electric drives, absolute power density values are low, of the order of 1 W/cm^3, which is up to three orders of magnitude lower than in piezoelectric stack actuators.

Other performance characteristics

The *temperature range of operation* for MS actuators is limited in practice by the material's Curie temperature. The Curie temperature for Terfenol-D is close to 380 °C, which is higher than the Curie temperature for lead zirconate titanate (PZT) materials.

Driving voltages for MS actuators are lower than those for piezoelectric stack actuators. Power supply for this technology is readily available from several manu-facturers. Unfortunately, the MS material itself is much less readily available. While PZT materials can be found in several grades, the only available MS material grade is Terfenol-D.

5.5.2 Scaling laws for magnetostriction

Scaling laws for magnetostrictive actuators are similar to scaling laws for Piezoelec-tric stack actuators. The force in MS actuators is proportional to its cross-sectional area and hence is proportional to the square of the dominant dimension, $F \propto L^2$.

Table 5.2 Operational characteristics and scaling trends for MS actuators.

Figures of merit	MS actuators
Force, F	High, $\geq 10^3$ N
Displacement, S	Up to 2000 ppm
Work density, W_V	$\approx 10^{-4} - 10^{-3}$ J/cm^3
Power density, P_V	≈ 1 W/cm^3
Bandwidth, f	$\geq 10^3$ Hz
Efficiency, η	Coupling factor 0.75
Scaling trends	
Force	$F \propto L^2$
Stroke	$S \propto L$
Work per cycle	$W \propto L^3$
Energy density	$W_V \propto L^0$
Bandwidth	$f \propto L^{-2}$
Power density	$P_V \propto L^{-2}$

The stroke is usually given as a percentage of the actuator's length; thus, since the actuator's dimensions are reduced linearly, so is the stroke, $S \propto L$.

The work per cycle can be readily derived from the force and stroke, that is, $W \propto L^3$. The ratio of work per cycle to the actuator volume is the energy density, which scales as $W_V \propto L^0$.

The time constant of an MS actuator can be estimated from the following expression:

$$[F] = [m][a] = [L^3]\frac{[L]}{[T^2]} \quad \Rightarrow \quad \tau \propto L^2 \qquad (5.27)$$

Since frequency bandwidth and time constant are inversely proportional, it follows that $f \propto L^{-2}$. As a result, the power density scales as $P_V \propto L^{-2}$.

Table 5.2 shows a summary of the most important figures of merit and scaling laws. A more detailed comparison with other emerging actuator technologies and traditional drives can be found in Chapter 7.

5.6   Applications

Case study 5.1:   Active vibration control of helicopter blades based on MS actuators

It is recognized that the full development of helicopter aviation is seriously limited by the inherent high vibration levels. The vibration is due mostly to the interaction of each blade with the wake of preceding blades. The problem of high vibration levels in helicopters is noise and to a great extent the limitation they impose on speed and load capacity. Vibration is also an important problem in that it causes accelerated pilot and passenger fatigue and, in some instances payload damage and frequent maintenance requirements (Fenn *et al.* (1996)).

Active vibration control of helicopter seems a promising strategy for improvement of vibration levels. In this context, conventional actuation mechanisms based on hydraulic and electric motors are less attractive than approaches based on certain emerging actuator technologies. This Case Study examines the application of Magnetostrictive actuators to active vibration control of helicopter blades, which may be regarded as one of the paradigmatic applications of MS actuators.

An individual blade control (IBC) system includes all the components usually found in motion control systems: sensors, electronics, controls and actuators. All the technologies required for IBC are well developed and ready for implementation. However, this is not true of actuators. In general, IBC systems can be classified into two categories: continuous active control of blade twisting (see Figure 5.15b) and discrete actuation of a servo-controlled surface (typically a flap or the blade tip; see Figure 5.15b).

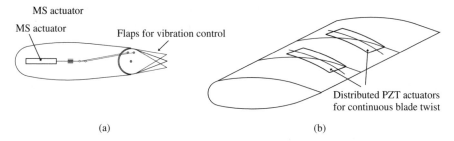

Figure 5.15 Concepts for vibration control in helicopter blades: (a) servo-controlled flaps and (b) continuous active control of blade twisting.

In continuous active control of blade twisting, the smart structure concept is commonly used. In this case, distributed induced-strain actuators produce continuous blade deformation in response to control actions. In the case of discrete actuation, flaps are used to generate localized aerodynamic forces.

The smart structure approach to blade vibration control has been implemented using PZT- distributed piezoelectric ceramics actuators, lead magnesium niobate (PMN) electrostrictive actuators and Terfenol-D magnetostrictive actuators. However, it has been shown (Fenn *et al.* (1996)) that approaches based on continuous blade twisting impose power requirements several orders of magnitude higher than discrete active control systems.

The particular requirements on actuators to achieve discrete (flap-based) vibration control of helicopter blades depend on innumerable factors, in particular, the ratio of blade length to width and the length of the control flaps. In general, MS actuators meet the actuation requirements for this application. In addition, MS actuators provide all-electric operation (thus simplified requirements as compared to hybrid – e.g., hydraulic – actuators), low mass, low voltage, insensitivity to centripetal acceleration, and simplicity and reliability.

When implemented in flap-based blade vibration control approaches, MS actuators can provide the required angular displacement of the flap with very compact solutions.

Case study 5.2: Prototype magnetic shape memory actuators

A new family of magnetostrictive materials, in which magnetostriction is twin induced, has recently been emerging as a highly promising technology. These are known as ferromagnetic shape memory alloys or magnetic shape memory alloys (MSMAs) and were introduced in Chapter 1.

In the first place, since they exhibit twin-induced strain, the stroke level is much higher (equivalent to the stroke level in thermally triggered SMAs) than in traditional Joule effect–based magnetostrictive materials. In the second place, since the martensitic transformation in MSMA actuators is magnetically triggered, they present a much faster response than thermally triggered SMAs.

Consequently, MSMAs share the good stroke characteristics of SMA actuators and the fast operating range of MS actuators. The technology is currently being explored in the areas of materials research and of operation and application.

One possible limiting factor in the development of the technology is the low availability of the material. Currently, MSMAs are only commercialized by Adaptamat Ltd., a company located in Finland (http://www.adaptamat.com) that markets material samples for research activities and is currently developing the first prototype actuators based on this technology.

This case study describes the first steps in the development of prototype MSMA actuators by Adaptamat Ltd. Here we discuss the development of three actuator prototypes spanning various force, stroke and operational frequency ranges. The information and pictures in this section are courtesy of E. Pagounis, Adaptamat Ltd.

The first prototype is the A5–2 actuator model from Adaptamat (see Figure 5.16). This is a high-stroke (maximum displacement of 3 mm) and low-force (blocked force in the order of 3 N) actuator. The basic configuration of MSMA actuators was discussed in Chapter 1. Basically, it consists of a MSMA stick aligned with the actuator direction. The magnetic field is set up perpendicular to the actuation direction. This triggers the reorientation of twin variants and initiates the actuation process.

In the case of the A5–2 actuator, the active MSMA stick comprises three active sections of $0.52 \times 2.4 \times 28$ mm so as to achieve a large overall stroke. The magnetic field is set up in this prototype by means of two coils, which can be electrically connected in parallel or in series. The actuator provides a fast drive with an electrical winding time constant of the order of 5 ms and a maximum operational frequency of 300 Hz.

The second prototype in this section is the small-stroke (maximum displacement 0.6 mm at 200 Hz), low-force (3 N) A06–3 actuator (see Figure 5.17). The basic configuration is similar to the previous case. Again, two coils are used to set up the magnetic field. This time a single MSMA stick ($0.55 \times 2.2 \times 20$ mm, with an active length of 15 mm) is used, resulting in a lower stroke.

The A06–3 prototype gives a faster drive than the previous example. The maximum attainable operational frequency is 600 Hz. It is also a much smaller

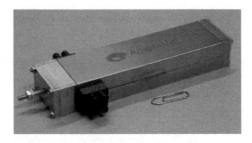

Figure 5.16   A5–2 MSMA actuator prototype from Adaptamat Ltd. (Photograph courtesy of E. Pagounis, Adaptamat Ltd.)

Figure 5.17   A06–3 MSMA actuator prototype from Adaptamat Ltd. (Photograph courtesy of E. Pagounis, Adaptamat Ltd.)

Figure 5.18   A1–2000 MSMA actuator prototype from Adaptamat Ltd. (Photograph courtesy of E. Pagounis, Adaptamat Ltd.)

actuator, weighing approximately 50 g as compared to the A5–2 prototype (which weighs around 500 g).

The following example illustrates the development of high-force, low-stroke actuators based on MSM materials: the A1–2000 actuator prototype from Adaptamat Ltd (see Figure 5.18). One of the current limiting factors in MSMAs is an intrinsically low Young's modulus (approximately three times lower than the Young's modulus of Terfenol-D) and high required magnetic fields (approximately twice those required in Terfenol-D). It is thought that this could be a serious impediment to the development of high-force MSMA actuators (since high forces require active material with a high cross-sectional area, which in turn requires a highly focused magnetic field over a large volume).

The A1–2000 prototype is an illustration example of an MSMA-based high-force actuator. In order to attain the high-force level (2 kN), 48 MSMA sticks are used mechanically in parallel. The dimensions of each stick are approximately 2.5 × 5.0 × 30 mm (with an active length of 25 mm, which means that the working strain is of the order of 40,000 ppm). Each stick has an active cross-sectional area of 12.5 mm^2, which makes a total active area of 600 mm^2 for the actuator.

The actuator's very large dimensions result make for an overall weight of 31 kg. Consequently, the operational frequency for this high-force drive is of the order of 100 Hz (and hence much lower than the previous examples, as one would expect).

These examples illustrate the development of MSMA actuators. They are claimed to be applicable to the field of fluidics (for the implementation of valves, pumps and injectors), vibration analysis and control (e.g. shakers, vibration dampers and sonar transducers), and as positioning devices in robotics, manipulators and linear drives in general.

6

Electro- and magnetorheological actuators (ERFs, MRFs)

Electro- and magnetorheological fluid (ERF and MRF) actuators are the only semi-active technology described in this book. They are essentially different from all the other technologies in that they can only (actively) dissipate the energy of the plant they are coupled to. In fact, they are, in a sense, energy sinks.

This feature makes them better suited for semiactive vibration cancellation applications. Like magnetostrictive (MS) actuators (see Chapter 5), these actuators are almost exclusively applied in vibration control applications, which is the main reason these two technologies are analyzed in the context of semiactive and active vibration control, respectively.

Another aspect common to MRF and MS actuators is transduction from the magnetic to the mechanical domain. For a sound and comprehensive discussion of these technologies, the concept of mechatronics must be broadened to take in aspects related to magnetic field setup and focusing concepts.

This chapter is organized to reflect such a broad conception of mechatronics. It begins by introducing some basic concepts of rheology, rheological materials, field-responsive fluids, electro- and magnetorheology and a historical note. This will provide the foundation for a full understanding of the design and control requirements of this technology. Readers interested in a brief introduction to the basics of magnetic properties of materials are referred to Section 5.1 in the previous chapter.

This brief introduction is followed by a detailed analysis of the mechatronic design concepts of ERF and MRF actuators. In particular, we pay special attention

Emerging Actuator Technologies: A Micromechatronic Approach   J. L. Pons
© 2005 John Wiley & Sons, Ltd

to the various different design concepts (flow, shear and squeeze modes). Tips on dimensioning are given for each concept according to application-related specifications. Also included are analyses of electronic and electrical aspects of driving circuits and the fundamentals of magnetic circuit design.

The section on control of ER and MR fluid actuators deals extensively with semiactive control of vibration. The basics of this control technique are analyzed and two classic control schemes, sky-hook and relative control, are described.

The control section analyzes performance characteristics for these technologies. Being semiactive, ER and MR actuators require a conceptually different actuation principle, so the performance figures of merit are essentially different from the ones used for other actuators. This aspect is addressed in Section 6.4.

Finally, the last section of this chapter is devoted to applications. Three case studies are presented, exemplifying the use of ERF and MRF actuators in the field of human lower limb prosthetics (the only application not strictly limited to vibration control), the application of this technology to cancellation of pathological tremor and its use in vibration control of driver's seats in heavy goods vehicles.

6.1   Active rheology: transducing materials

According to the British Society of Rheology, the term rheology defines the science of the deformation and flow of matter. In principle, it could include the study of all materials that deform or flow but, by convention, classical (obeying the Hooke's law) elastic solids and viscous (Newtonian) liquids are excluded. Both non-Newtonian and elastic liquids and viscoelastic solids are included in the science.

This chapter will focus on the so-called electro- and magnetorheological fluids. The rheological behavior of these alters in an interesting way when they are subjected, respectively, to electric and magnetic fields. But, let us first introduce some basic concepts of rheology.

6.1.1   Basics of rheology

Rheology, then, is the science that studies the elastic, plastic and viscous properties of materials. Rheology is particularly interested in describing the stress–strain relationship in flowing fluids. The mathematical expressions formulating the stress–strain relationship in fluids are known as *rheological equations of state*.

Fluids are divided into two broad categories: Newtonian and non-Newtonian.

1. *Newtonian fluids*. In Newtonian fluids, rheological equations of state describe a linear relationship between shear stress and rate of shear as in the following expression:

$$\tau = \eta\dot{\gamma} \tag{6.1}$$

In Equation 6.1, τ is the shear stress in the fluid, γ is the shear strain and η is the constant of proportionality known as *viscosity*. It is worth noting that for Newtonian fluids the viscosity is constant and independent of time or applied shear stress.

2. *Non-Newtonian fluids*. Within this category, all the remaining fluids are classified. Roughly, they can be, in turn, included in one of the following groups:

   (a) *Time-independent fluids*. Here, the shear rate is only a function of shear stress, that is, τ is a function of γ.

   (b) *Time-dependent fluids*. Rheological equations of state depend on how the fluid has been sheared and on previous history.

   (c) *Elastoviscous fluids*. These fluids are predominantly viscous but exhibit partial elastic recovery after deformation.

As discussed in detail in the following sections, the rheological characteristics of electro- and magnetorheological fluids can be classified as time-independent fluids. Time-independent fluids can be further analyzed according to the specific form of the rheological equation of state that describes their behavior. In general, the mathematical formulation of the stress–strain behavior in time- independent fluids is similar to Equation 6.2

$$\tau = \tau_y + \eta \dot{\gamma}^n \qquad (6.2)$$

where, τ, η and γ are as defined above, τ_y is the yield stress and n describes the functional relationship between stress and rate of strain.

In establishing an analogy between Newtonian and time-independent fluids, the following expression characterizes the *apparent viscosity* in time-independent fluids:

$$\eta_{ap} = \frac{\tau}{\dot{\gamma}^n} = \frac{\tau_y}{\dot{\gamma}^n} + \eta \qquad (6.3)$$

Depending on the particular structure of the rheological equation of state, Equation 6.2, the following cases can be found in time-independent fluids:

- *Bingham Fluids*. A Bingham fluid is an idealized material whose internal structure collapses above a yield stress τ_y. For higher stresses, the rheological behavior is linear. In Bingham fluids, the rheological equation of state reduces to:

$$\tau = \tau_y + \eta \dot{\gamma} \qquad (6.4)$$

Electro- and magnetorheological fluids are commonly considered Bingham fluids in which the yield stress is a function of the electric field, E, or magnetic field, H, $\tau_y = \tau_y(E)$ or $\tau_y = \tau_y(H)$, respectively.

- *Pseudoplastics*. These fluids exhibit no yield stress, $\tau_y = 0$, but the stress, τ, increases less than linearly with the rate of strain, $\dot{\gamma}$. This type of fluid is well represented by the equation:

$$\tau = \eta \dot{\gamma}^n \qquad (6.5)$$

where η and n are constants, and in addition $n < 1$. As the shear stress increases, the slope of the stress versus rate of strain relationship decreases: that is, the apparent viscosity gets less as mixing increases. These fluids are also known as *shear-thinning* fluids.

- *Dilatant Fluids*. These fluids follow the same rheological equation of state as in the previous case (Equation 6.5), but in this case the following inequality holds, $n > 1$. As a result, the curve for a dilatant fluid looks the same as for a pseudoplastic fluid, except that the slope of the stress–rate of strain relationship increases. Consequently, the apparent viscosity increases along with the shear rate. These fluids are also known as *shear-thickening* fluids.

The rheological equation of state for all the above-described fluids is represented graphically in Figure 6.1. It is apparent that the behavior of some types of fluids can closely resemble that of other fluids. This is true of shear-thinning fluids, whose behavior at a low applied rate of stress is similar to that of an idealized

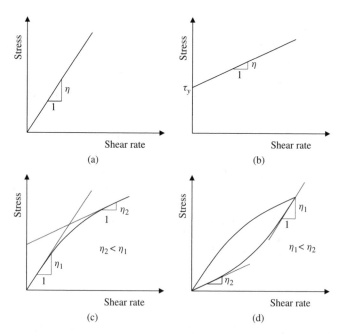

Figure 6.1   Stress versus shear rate relationship for Newtonian (a) and non-Newtonian, Time-independent fluids (b, c and d).

Newtonian fluid. In the case of higher mixing rates, a linear approximation similar to Bingham fluids is possible.

6.1.2 Field-responsive fluids

In general, we may speak of three main field-responsive fluids, namely, magnetorheological fluids (MRF), electrorheological fluids (ERF) and ferrofluids. Field-responsive fluids typically undergo abrupt changes in their rheological, magnetic, electric, thermal, optical, acoustic, mechanical and physical properties in general when subjected to an external electric or magnetic field.

Although the practical study and application of field-responsive fluids is relatively new (the last two decades), their electromagnetic properties have been known since the late-1940s. Electrorheology and magnetorheology, in particular, were discovered by Winslow (1949) and Rabinow (1948) almost simultaneously. The following paragraphs briefly analyze the three field-responsive fluids and compare their rheological properties.

Magnetorheological fluids

Magnetorheological fluids are suspensions of noncolloidal, multidomain and magnetically soft particles in base organic or aqueous liquids. The size of the magnetic particles suspended in the base liquid may vary from 0.1 to 10 μm.

Magnetorheological fluids behave similarly to Bingham plastics. When not subjected to any external magnetic field, the apparent viscosity of MR fluids is of the order of 0.1 to 10 Pa·s. Where an external magnetic field is applied, within a very short response time, of the order of milliseconds, the apparent viscosity will undergo an abrupt change of several (up to 5–6) orders of magnitude.

Moreover, the yield stress of MR fluids (according to the Bingham model) is also modified. Depending on the type of MR fluid and the applied magnetic field, yield stresses up to ≈100 kPa have been reported. The maximum applicable magnetic field is limited by saturation to values of ≈250 kA/m.

The fraction concentration of magnetic particles in the base liquid can vary amongst different MR fluids. Typical values for the fraction concentration are of the order of 20 to 60%. It is believed that the increased yield stress and apparent viscosity is due to the build up of magnetic particle chains in response to the interparticle magnetic interaction.

Where a low shear stress is applied to the MR fluid, particle chains will first be stretched and deformed. If the applied stress is increased further, chains will eventually break and the material will start to flow.

Electrorheological fluids

Electrorheological (ER) fluids are suspensions of electrically active particles in an electrically insulating liquid. Particle sizes in ER fluids are, on average, one order of magnitude larger than particle sizes in MR fluids (between 1–100 μm).

Similar to MR fluids, the electrically active particles in ER fluids are subject to interparticle interactions causing the formation of particle chains. Again, ER fluids conform to a Bingham model, but the maximum yield stress is 2 orders of magnitude lower in ER fluids than in MR fluids.

Maximum yield stress values are typically of the order of 1 to 5 kPa. The maximum applicable electric field in ER fluids is limited by the breakdown of the dielectric characteristics of the fluid.

Ferrofluids

Ferrofluids are colloidal suspensions of single-domain magnetic particles in either aqueous on nonaqueous liquids. The typical particle size in ferrofluids is of the order of a few (commonly 5 to 10) nanometers.

Ferrofluids undergo field-induced viscosity changes but do not present yield stress. This is basically because the magnetic particles are ultrasmall: thermal agitation of the particles produces Brownian forces that prevent the formation of interparticle chains.

The application of an external magnetic field generates body forces in the ferrofluid, which in turn cause changes in the viscosity (which can double because of this effect). Such fluids are unsuitable for use as actuator technologies owing to the high cost of achieving only slight modification of rheological properties.

The various different properties of field-responsive fluids are summarized in Table 6.1.

6.1.3 Electro- and magnetorheology

As explained in the previous section, the most interesting change in rheological properties (for the application of field-responsive fluids to actuators) occurs in ER and MR fluids. In this section, the discussion focuses on an analysis of the rheology of ER and MR fluids to provide an insight into the way ER and MR fluids are used as actuators.

The *electrorheological effect (ERE)* and the *magnetorheological effect (MRE)* can be defined respectively as the increase in shear stress because of an applied electric and magnetic field.

Mechanical operation with ER or MR fluids can be analyzed with respect to two situations: *preyield* and *postyield operation*. Both situations depend on the level of mechanical stress being exerted on the fluid. Where the shear stress is below the elastic-limit yield stress, $\tau \leq \tau_e$, the fluid behaves as a solid. Therefore, in this condition, the shear rate is zero, $\dot{\gamma} = 0$, and shear strain and stress are proportional. The constant of proportionality is the shear modulus, G, of the fluid:

$$\tau = G\gamma, \ \dot{\gamma} = 0 \qquad \tau < \tau_e \qquad (6.6)$$

Where the applied shear stress is higher than the static yield stress, $\tau \geq \tau_s$, the material flows and the shear stress is proportional to the rate of shear:

$$\tau = \tau_y + \eta\dot{\gamma} \qquad \tau \geq \tau_s \qquad (6.7)$$

Table 6.1 Comparative analysis of properties of MR fluids, ER fluids and ferrofluids.

Property	Units	MR Fluids	ER Fluids	Ferrofluids
			Physical properties and constituents	
Particle type	–	Fe, ferrites	Zeolites, polymers, $B_aT_iO_3$	Ceramics, Fe, ferrites
Particle size	μm	0.05–10	0.5–100	$2-10 \cdot 10^{-3}$
Base liquid	–	nonpolar oils, polar liquids	Oils	Oils, water
Density	kg/m^3	$3-5 \cdot 10^3$	$1-2 \cdot 10^3$	$1-2 \cdot 10^3$
Max. yield stress	kPa	50–100	2–5	–
Operating temp.	C	−40 to +150	−25 to +125	–
No field viscosity	mPa·s	100–1000	100–1000	2–500
			Electromagnetic properties	
Maximum field	–	≈250 kA/m	≈4 kV/mm	–
Field-induced change	kPa	$\tau_y(H) \approx 10^2$	$\tau_y(H) \approx 5$	$\dfrac{\eta(H)}{\eta(0)} \approx 2$
Max. energy density	kJ/m^3	100	1	–
Response time	s	$<10^{-3}$	$<10^{-3}$	–

The *static viscosity*, η, of ER and MR fluids is essentially determined by the viscous characteristics of the base fluid. However, both the yield stress and the shear modulus are functions of the applied electric or magnetic field. For practical purposes, linear models are used to predict the variation of yield stress and shear modulus in response to the field. According to these models, both the yield stress and the shear modulus are quadratic functions of the applied field. For the particular case of MR fluids, this becomes

$$\tau_y = \tau_y(H) \propto H^2 \tag{6.8}$$

$$G = G(H) \propto H^2 \tag{6.9}$$

Figure 6.2a and b shows the shear stress versus shear strain relationship and the stress versus rate of strain relationship. Given this mechanical behavior of ER and MR fluids, they may be actuated as either viscoelastic or plastic materials. Likewise, either preyield or postyield operation is used, depending on the type of application.

A closer look at the shear stress–strain relationship for field-responsive fluids shows that three different yield stresses can be defined. The minimum stress required to make the fluid flow is known as the *static yield stress*, τ_s. The maximum stress defining the limit of the elastic behavior of MR and ER fluids is known as *elastic-limit yield stress*, τ_e. The *dynamic yield stress*, τ_y, is the yield stress defined by the Bingham model of the MR and ER fluid and corresponds to the yield stress in the transition from liquid to solid behavior. The three different yield stresses are depicted in Figure 6.3.

Using experimental data, it can be shown that both static and dynamic yield stresses occur at different levels. Therefore, the behavior in a cyclic transition from liquid to solid and back is hysteretic. The relative magnitudes of static and dynamic yield stress depend on the particular ER and MR fluid. However, it is most common for the static yield stress to be higher than the dynamic yield stress: $\tau_s > \tau_y$.

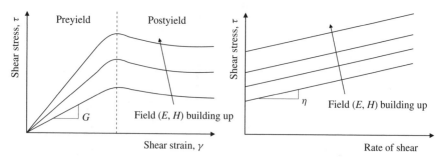

Figure 6.2 Qualitative representation of the field-dependent yield shear modulus (a) and field-dependent yield stress (b) in ER and MR fluids.

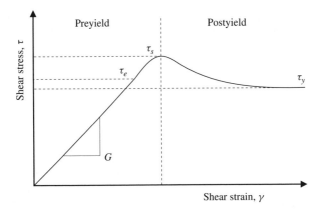

Figure 6.3 Relative position of elastic-limit, static and dynamic yield stresses in MR and ER fluids.

6.2 Mechatronic design concepts

6.2.1 Shear, flow and squeeze modes

ER and MR actuators are devices combining concepts from geometrical transducers and transducing materials. The main purpose of the geometry of MR or ER devices is to determine the manner in which the fluid is sheared. When dealing with MR or ER actuators, there are three common device geometries or shearing modes: first, the shear or clutch mode; second, the flow or valve mode and third, the squeeze mode.

The first category is commonly used in the design of clutches or brakes. When the design object is a damper, on the other hand, it is usually the flow or the squeeze mode that is used. The main aspects of these actuation modes are briefly reviewed in the following paragraphs.

In general, the object of the design is to achieve a programmable relationship between the applied force or pressure drop and the resulting device velocity or the volumetric flow rate. This objective relationship will take the following form:

$$X(E) = X_\eta(Y) + X_\tau(E) + X_I \qquad (6.10)$$

$$X(H) = X_\eta(Y) + X_\tau(H) + X_I \qquad (6.11)$$

where Equation 6.10 is to be considered for ER devices and Equation 6.11 for MR devices. $X(E)$ and $X(H)$ represent the electric or magnetic field–dependent force or pressure drop. Y represents the device's velocity or volumetric flow rate. $X_\eta(Y)$ is the field-independent force or pressure drop due to the static viscosity, which is a function of Y. Finally, $X_\tau(E)$ and $X_\tau(H)$ are field-dependent forces or pressure drops due to the programmable yield stress, τ_y, and X_I is the device force or pressure drop corresponding to fluid inertia.

The term X_I can generally be ignored in quasistatic situations, and only the viscosity and the yield stress terms will be left. In these devices, the ratio of the field-dependent term at the maximum admissible field, $X_\tau(H_{max})$ or $X_\tau(E_{max})$, to the viscosity term, $X_\eta(Y)$, is known as the *mechanical control ratio*, λ:

$$\lambda_E = \frac{X_\tau(E_{max})}{X_\eta(Y)} \tag{6.12}$$

$$\lambda_H = \frac{X_\tau(H_{max})}{X_\eta(Y)} \tag{6.13}$$

The particular formulation of the control ratio, the field-dependent force and the viscosity terms depends on the device geometry and is discussed briefly in the following sections. The mechanical control ratio is an indication of the ratio of the dissipative force in the 'ON' state to the dissipative force in the 'OFF' state. The design objective is typically to achieve the maximum dissipation in the 'ON' state while allowing minimum dissipation when there is no external field.

Shear or clutch mode devices

In this design concept, the fluid is confined between two plates. One of the plates is considered fixed, the other is subject to a relative displacement, and the distance between them is constant. For a schematic illustration of this design concept, see Figure 6.4.

In the figure, the lower plate is fixed. A force, $X(E) = F(E)$, is applied to the upper plate in the horizontal plane so that the relative displacement occurs at a speed $Y = \dot{x}$, while the fluid is subject to shear. For the particular configuration shown in Figure 6.4, according to Phillips (1969), the field-independent viscosity

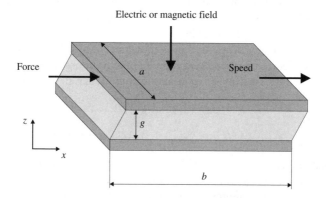

Figure 6.4 Schematic representation of the geometry of shear mode devices: the electric or magnetic field is applied perpendicular to the shearing plates; the applied force produces a shear stress at the fluid. (Reproduced by permission of Lord Corporation.)

term, F_η, and the field-dependent yield stress term, $F_\tau(E)$, are

$$F_\eta = \frac{\eta \dot{x} b a}{g} \tag{6.14}$$

and

$$F_\tau(E, H) = \tau_y(E, H)ba \tag{6.15}$$

The mechanical control ratio, λ_E or λ_H, for the shear mode configuration can be obtained by simply taking the ratio of Equations 6.15 to 6.14.

Flow or valve mode devices

The configuration for this design concept is shown schematically in Figure 6.5. In this case, the active fluid is again confined between two plates. Both plates are fixed in this configuration, so that a pressure drop, $X(E) = P(E)$, forces the fluid to flow at a volumetric flow rate, $Y = Q$.

In the case of ER fluids, an electric field, E, is generated across the interplate gap. Where MR fluids are concerned, a magnetic circuit is designed with the fluid-active area between the plates included in its path. In this configuration, according to Phillips (1969), the field-dependent pressure drop, $P_\tau(E)$ or $P_\tau(H)$, and the field-independent viscous term, $P_\eta(Q)$, across the device can be calculated thus:

$$P_\eta(S) = \frac{12\eta Q b}{g^3 a} \tag{6.16}$$

and

$$P_\tau(E, H) = \frac{c\tau_y(E, H)b}{g} \tag{6.17}$$

Again, the mechanical control ratio can be derived from Equations 6.16 and 6.17.

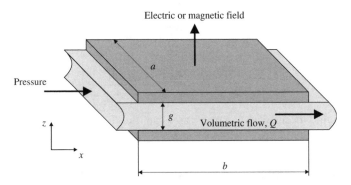

Figure 6.5   Schematic representation of the geometry of flow mode devices: the electric or magnetic field is applied perpendicular to the fixed plates, the applied pressure causes the ER or MR fluid to flow between the plates. (Reproduced by permission of Lord Corporation.)

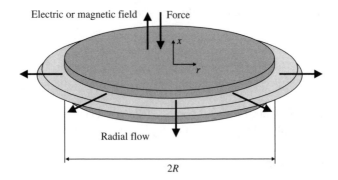

Figure 6.6 Schematic representation of the geometry of squeeze mode devices: the electric or magnetic field is applied perpendicular to the squeezing plates; the applied force is collinear to the field and produces a radial flow.

Squeeze mode devices

The design concept for squeeze mode devices is the only one that exploits the relative perpendicular displacement of the plates. This configuration is illustrated schematically in Figure 6.6.

A force, $F(E)$, is applied perpendicularly to the plates. The fluid confined between the two plates is forced to flow horizontally out of the interplate gap. The resulting relative velocity of the plates will thus be $\dot{x}$. In this configuration, a nominal gap distance, g_0, is considered.

According to Jolly and Carlson (1996), the velocity-dependent force due to viscosity, $F_\eta(\dot{x})$, and the field-dependent yield stress term, $F_\tau(E)$ or $F_\tau(H)$, can be calculated thus:

$$F_\eta = \frac{3\pi\eta R^4 \dot{x}}{2(g_0 - x)^3} \tag{6.18}$$

and

$$F_\tau(E, H) = \frac{4\pi\tau_y(E, H)R^3}{3(g_0 - x)} \tag{6.19}$$

Again, the mechanical control ratio can be derived from Equations 6.18 and 6.19.

6.2.2 Device dimensions according to specifications

In the design of semiactive energy dissipation actuators based on ER or MR fluids for a particular application, the required dissipation power, P_m, and the required maximum dissipative force, F_τ, are usually application-related inputs. Equations 6.14 through 6.19 can be manipulated to give the following expression for the required volume of active fluid:

$$V = K \frac{\eta}{\tau_y^2} \lambda P_m \tag{6.20}$$

where K is a mode-dependent constant, λ is the application requirement in terms of mechanical control ratio and P_m is the mechanical power that is required to be dissipated.

In Equation 6.20, P_m can be obtained from the application specifications in terms of maximum dissipative force and maximum velocity or flow rate, that is, $P_m = F_\tau \cdot \dot{x}$ or $P_m = P_\tau \cdot Q$. Also, in Equation 6.20, λ is specified according to the ratio of maximum force or pressure to acceptable minimum force or pressure (see Equations 6.12 or 6.13).

Table 6.2 summarizes the different design equations for all three concepts. A possible design methodology would comprise the following steps:

1. Find the required power to be dissipated, P_m, from the application specifications, maximum force (pressure), $X_\tau(E_{\max}, H_{\max})$, and velocity (volumetric flow rate), Y.

2. Find the acceptable force (pressure), $X_\eta(\dot{x})$, in the 'OFF' state.

3. Select the fluid characteristics, η and τ_y, from data sheets of available fluids.

4. Use Equation 6.20 to trade off device geometry, possibly by taking into account other device geometry constraints from the target application.

Flow mode and squeeze mode devices are generally used for semiactive damping applications. The former is suitable for low-force, high-displacement applications, while the latter is usually more appropriate for low-displacement, high-force vibration damping applications.

6.2.3 Driving electronics for ER and MR devices

The electrical equivalent circuit for a magnetorheologic damper can best be approximated by the coil circuit responsible for setting up the magnetic field across the fluid-active area. As such, it is represented by a resistor and an inductance in series (see Figure 6.7a). The resistor takes into account the resistive losses at the coil, while the inductance represents the electrical load of the coil.

The electrical equivalent circuit for an electrorheologic device is best approximated by a resistor and a capacitor in parallel (see Figure 6.7b). Both the resistor and the capacitor will generally exhibit variable characteristics during operation. This is apparent in the case of a damper working in the squeeze mode. The distance between electrodes will be subject to continuous changes, which, in turn, will determine variable capacitance.

The following paragraphs briefly introduce a few design considerations.

Table 6.2 Summary of design formula for shear, flow and squeeze mode device configurations.

Device geometry	Viscous force, $F_\eta(\dot{x})$	Dynamic force, $F_\tau(E,H)$	Active volume, V
Clutch or shear mode	$\dfrac{\eta \dot{x} b a}{g}$	$\tau_y(E,H)ba$	$V = K\dfrac{\eta}{\tau_y^2}\lambda P_m,\quad K=1$
Squeeze mode	$\dfrac{3\pi \eta R^4 \dot{x}}{2(g-x)^3}$	$\dfrac{4\pi \tau_y(E,H)R^3}{3(g-x)}$	$V = K\dfrac{\eta}{\tau_y^2}\lambda P_m,\quad K=\dfrac{12}{c^2}$
Valve or flow mode	$\dfrac{12\eta Q b}{g^3 a}$	$\dfrac{c\tau_y(E,H)b}{g}$	$V = K\dfrac{\eta}{\tau_y^2}\lambda P_m,\quad K=\dfrac{32}{27}$

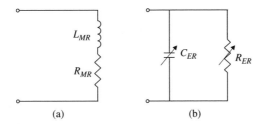

(a)   (b)

Figure 6.7   Equivalent electrical circuit for ER and MR fluid loads: (a) the ER
load is equivalent to the parallel connection of a variable capacitor and resistor,
(b) the MR load is best approximated by the series connection of an inductance
and a resistor.

Power amplifiers for MR devices

Magnetorheologic actuators are generally preferred to their electrorheologic counterparts. The main reasons for this are as follows:

1. The electrical equivalent circuit parameters for MR devices are much more
   stable than the parameters for ER devices.

2. Low-voltage, high-current power sources are required, and these are generally readily available.

3. Although MR actuators are slightly larger than ER actuators (mainly due
   to the coils that set up the magnetic field), in terms of the overall device
   (including power source and control electronics), MR devices offer more
   compact solutions.

   In designing power drivers for MR devices, overvoltage protection must be
included in the electronic control circuit. Since the MR fluid is an inductive load,
according to Faraday's law of induction, the voltage across the coil will be defined
by the expression:

$$V(t) = -L\frac{\mathrm{d}i}{\mathrm{d}t} \tag{6.21}$$

   If for any reason there is a break in the connection between the load and
the power amplifier, the current drop (high $\mathrm{d}i/\mathrm{d}t$) will produce an unacceptably
high voltage.

   According to Yang (2001), the coil can be protected from this high voltage
by means of a transient voltage suppressor (TVS) (see inset in Figure 6.8). The
TVS represents a high resistance (virtually an open circuit condition) during normal
operation. In the event of sudden high voltages, when a voltage threshold is reached,
it starts conducting and clamps the voltage to an acceptable level. The current is
then dissipated through heat at the coil resistance.

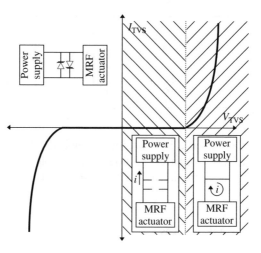

Figure 6.8   Typical operation curves of a bipolar TVS. Inset (bottom right) the equivalent operation below and above threshold voltage, open circuit and short circuit respectively.

Power amplifiers for ER devices

The electrical load represented by electrorheologic actuators is highly nonlinear. It is strongly dependent on temperature, electrical field and the volumetric flow in the actuator. In addition, the electrical characteristics will typically be modified by the mechanical interaction with the output load. In general, we may say (Stiebel and Janocha (2000)) the following:

1. The equivalent resistor in Figure 6.7a is a function of the ER fluid's conductivity. The conductivity of ER fluids doubles with each temperature rise of 6°.

2. The capacitance in Figure 6.7a is influenced by the geometric changes in the actuator. In squeeze and shear mode devices, the capacitance will depend on the electrode gap and the effective area, respectively.

3. The output mechanical load will generally produce a reaction on the electrical part.

4. If no means are provided for sinking current by the power electronic circuit, the time response of the ER device will be influenced by the discharging time of the capacitance in Figure 6.7a. It can be demonstrated that the discharge time in these conditions will depend only on fluid properties (conductivity and permeability) and not on device geometry. Because conductivity is strongly dependent on temperature, there can be differences of up to 3 orders of magnitude in the discharging time.

In order to overcome these nonlinear phenomena, two- or four-quadrant power amplifiers are preferred. One-quadrant amplifiers can only source current. If one-quadrant amplifiers are used, the discharge of the capacitor must be fully met by the conductance on the fluid. Readers are referred to Stiebel and Janocha (2000) for an excellent description of driving conditions for ER devices.

6.2.4 Design of magnetic circuits in MR devices

In general, the design criteria described above for MR actuators will produce an operating point close to magnetic saturation in the fluid: in other words, the MR fluid's properties will be optimally utilized.

In magnetic circuit design, an analogy with electrical circuits is established. The effort in the magnetic circuit corresponds to the magnetomotive force, $\mathcal{F}$, and is analogous to the voltage in electrical circuits. The flow in magnetic circuits is the magnetic flux, Φ, which is analogous to the electrical current, i. The analogue of the electrical resistance in purely resistive electrical circuits is the reluctance, $\mathcal{R}$.

Carrying on with the analogy between electrical and magnetic circuits, Kirchoff's Laws for electrical circuits also have their magnetic counterpart:

1. *Kirchoff's Flux Law (KFL)*. This is the analogue of Kirchoff's Current Law (KCL) for electrical circuits. It states the continuity of magnetic flux along the magnetic circuit and is the result of the application of Gauss's Law:

$$\oint_S \mathbf{B} \cdot \hat{\mathbf{n}} \, da = 0 \qquad (6.22)$$

For simple magnetic circuit geometries, Equation 6.22 reduces to the following expression:

$$\sum B_i \cdot A_i = 0 \qquad (6.23)$$

In practice, it states that the magnetic flux ($\Phi = BA$) along the circuit (and thus comprising both the magnetic guide and the fluid's active area) is continuous.

2. *Kirchoff's Magnetomotive Law (KML)*. This is the equivalent of Kirchoff's Voltage Law (KVL) for electrical circuits. It can be worked out from Ampere's Law:

$$\oint_C \mathbf{H} \, dl = \int_S J \hat{\mathbf{n}} \, da \qquad (6.24)$$

where C is a closed loop (the magnetic circuit) enclosing a planar area S, J is the electrical current density flowing through S and $\hat{\mathbf{n}}$ is a unit vector normal to S. Note that the second term in Equation 6.24 is the electrical current

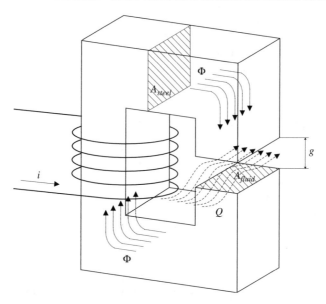

Figure 6.9   Reluctance circuit for focusing the magnetic flux into the active volume of MR fluid. (Reproduced by permission of Lord Corporation.)

flowing through S. Again, for simple circuit geometries, if a coil with n_i turns is used to establish the magnetic flux, Equation 6.24 reduces to

$$\sum_C H_i \cdot l_i = n_c i \qquad (6.25)$$

The magnetic flux will be set up at the coil, and a magnetic circuit will focus the magnetic flux into the active volume of the MR fluid (see Figure 6.9). Ideally, the magnetic flux is focused by providing maximum field energy in the fluid region while minimizing the magnetic energy losses in the reluctance circuit.

Using Kirchoff's laws for magnetic circuits, the process for designing the reluctance circuit for an MR application is typically as follows (Lord Corporation (1999)):

1. The fluid operating point will be determined by the pair (H_{fluid}, B_{fluid}) as given by data sheets. The total magnetic flux at the fluid area will be given by

$$\Phi_{fluid} = B_{fluid} \hat{A}_{fluid} \qquad (6.26)$$

where $\hat{A}_{fluid}$ is the effective flux area at the MR fluid. When the effective flux area is used, the fringing effect of the magnetic field must be taken into account (see Lord Corporation (1999)).

2. The flux density will be worked out along the reluctance circuit on the basis of the continuity of the magnetic flux.

3. The flux density in the magnetic guide (steel) is determined thus:

$$B_{steel} = \frac{B_{fluid} A_{fluid}}{A_{steel}} \tag{6.27}$$

Note that different sections of the steel circuit may be subject to different flux densities if the cross-sectional area is not constant along the magnetic circuit. B_f is used to determine the operating point along the steel, (H_{steel}, B_{steel}).

4. The number of coils, n_c, is determined according to the required magnetic field, H_{steel}, using Kirchoff's law for magnetic fluids:

$$n_c i = H_{fluid} L_{fluid} + H_{steel} L_{steel} \tag{6.28}$$

6.3 Control of ERF and MRF

Unlike all the other technologies described in this book, ER and MR actuators are semiactive devices. ER and MR dampers are able to dissipate mechanical energy, but they are not capable of providing energy to the plant they control.

Semiactive devices are essentially passive and are very much restricted to control forces opposing the relative velocity, Karnopp *et al.* (1974). Let us consider the damper fluid (ER or MR) shown schematically in the inset in Figure 6.10. The condition of dissipative forces reads as

$$f_{damper} \cdot (\dot{x}_1 - \dot{x}_2) \geq 0 \tag{6.29}$$

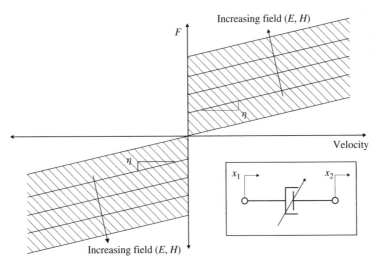

Figure 6.10 Representation of the operating region for ER–MR dampers. Schematic representation (inset) of a variable yield stress ER or MR damper.

The operating range for the field-responsive damper according to this admissible force is shown in Figure 6.10. The use of ER and MR devices will be closely connected with vibration and shock control applications. In the following paragraphs, the problem of vibration suppression by means of semiactive dampers is analyzed as the paradigmatic application of MR and ER devices.

The vibration suppression or isolation problem, which is illustrated in Figure 6.11, can be stated as follows:

Given a vibration source mechanically connected to a sensitive piece of equipment (or actor), use passive, active or semiactive means to reach a compromise between amplification at resonances and attenuation at high frequency.

Let us consider the second-order system depicted in the inset in Figure 6.11. The sensitive load M is mechanically connected to a vibration source, m, through a spring and a damper. According to systems theory, when the damping is very low, the mechanical amplification at resonance will become high and the high-frequency attenuation will asymptotically be -40 dB/decade. The passive solution to the vibration isolation problem consists in increasing the system damping so that the amplification at resonance is limited.

The active damping concept was analyzed in more detail in Chapter 5. Briefly, control can be approached by either *feed-forward* or *feedback* schemes. The former is used when there is a known model of the disturbance. The latter is more general in scope and can be used in both narrow- and broadband applications.

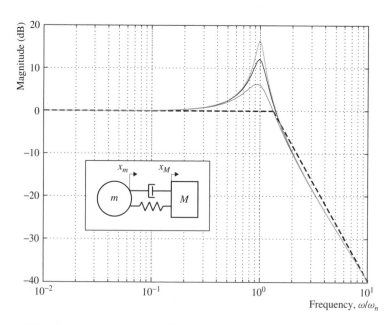

Figure 6.11   Design goal (broken line) for the vibration isolation problem of a second-order system subject to vibration sources (inset).

The design goal for vibration isolation systems is indicated by the broken line in Figure 6.11. Ideally, the system should filter out all frequency components below the corner frequency, $\omega = \sqrt{2}\,\omega_n$, while preserving the -40 dB/decade attenuation rate at high frequency.

The semiactive implementation of vibration isolation with MR and ER dampers is limited to the admissible operating region depicted in Figure 6.10. Given this limitation, in the following paragraphs, what are called the *sky-hook* vibration control scheme and the *relative* control approach are discussed as classic vibration control schemes subject to the constraint of semiactive operation.

6.3.1 Sky-hook vibration isolation

Consider a force generator, F_g, placed in parallel to a spring, both in the mechanical path connecting a vibration source, m, and a mass, M (see Figure 6.12). The force generator is driven in such a way that it provides a force proportional but opposite to the absolute velocity of M:

$$F_g = -Ds X_M \qquad (6.30)$$

where s is the Laplace variable, $s X_M$ is the absolute velocity of the mass and D is the gain.

This is equivalent conceptually to attaching the mass M to a fixed point in space (the sky) through a damper. In practice, this feedback concept can be implemented by measuring either the acceleration of the mass, $s^2 X_M$, or the force transmitted to it, F_M, by the force generator and the spring. The acceleration or the force is then passed through an integral controller, D/s. The sky-hook controller is depicted as a block diagram in Figure 6.13.

The open loop transfer function of the system of Figure 6.13 (Preumont and De Man (2004)) is

$$H(s)G(s) = \frac{D}{s}\frac{Ms^2 X_M}{F_g} = D\frac{mMs^2}{s(mMs^2 + k(M+m))} \qquad (6.31)$$

The open loop transfer function has one pole at the origin, $s_{p1} = 0$, and two pure imaginary poles at $s_{p2,p3} = \pm j\sqrt{k(m+M)/mM}$. In addition, it has two zeros

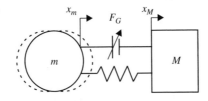

Figure 6.12 Scheme of a mass (M) being driven by undesired vibrations induced by a second mass m.

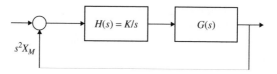

Figure 6.13   Block diagram of the sky-hook controller: integral control over the sprung mass acceleration provides the absolute velocity required for the sky-hook concept.

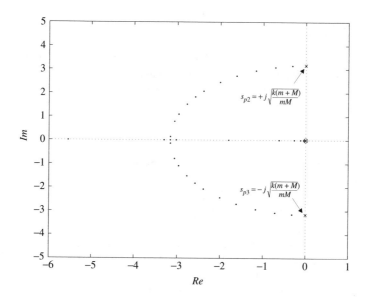

Figure 6.14   Root locus of the transmissibility between vibration source and sprung mass. The system is unconditionally stable for all feedback gains.

at the origin, $s_{z1,z2} = 0$, plus a third zero at $s = \infty$. A simple inspection of the root locus of the isolator clearly shows the unconditional stability of the sky-hook control scheme (see Figure 6.14).

Furthermore, the transmissibility between the vibration source and the mass (Preumont and De Man (2004)) is found to be

$$\frac{X_M}{X_m} = \frac{k}{Ms^2 + MDs + k} \tag{6.32}$$

thus exhibiting a −40 dB/decade attenuation rate for high frequency.

Unfortunately, a semiactive damper will not fully meet the requirements of the force generator, F_g, in the sky-hook control strategy. Instead, the control strategy for semiactive dampers is to set the control signal to the appropriate value wherever inequality 6.29 is fulfilled and let the control signal be zero when inequality is violated. The sky-hook control scheme for semiactive dampers is formulated thus:

$$f_{damper} = D\dot{x}_M \quad \text{if} \quad \dot{x}_M \cdot (\dot{x}_M - \dot{x}_m) \geq 0$$

$$f_{damper} = 0 \qquad \text{if} \quad \dot{x}_M \cdot (\dot{x}_M - \dot{x}_m) < 0 \qquad (6.33)$$

In general, the relationship between the applied force and the velocity in an ER or MR damper can be described by the following expression:

$$f_{damper} = f_\tau + f_\eta(\dot{x}_M - \dot{x}_m) \qquad (6.34)$$

where $f_\tau = f_\tau(H)$ for MR dampers and $f_\tau = f_\tau(E)$ for ER dampers, in general $f_\tau = f_\tau(u)$ (with u the control signal), and f_η is proportional to the relative velocity, $f_\eta = c \cdot |\dot{x}_M - \dot{x}_m|$. Equation 6.34 can be rewritten as follows:

$$f_{damper} = \left[\frac{f_\tau(u)}{|\dot{x}_M - \dot{x}_m|} + c \right] |\dot{x}_M - \dot{x}_m| \qquad (6.35)$$

Table 6.3   Conditions of activation in a semiactive sky-hook control scheme.

Equivalent diagram	Condition	Damper force, F_g
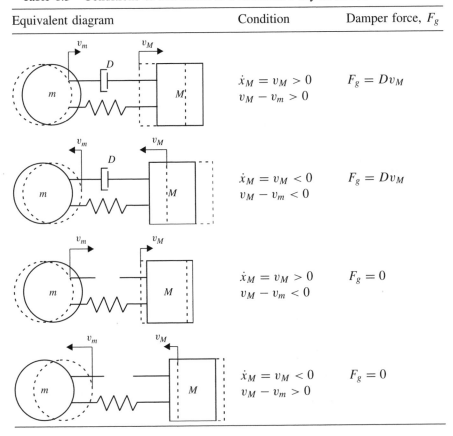	$\dot{x}_M = v_M > 0$ $v_M - v_m > 0$	$F_g = Dv_M$
	$\dot{x}_M = v_M < 0$ $v_M - v_m < 0$	$F_g = Dv_M$
	$\dot{x}_M = v_M > 0$ $v_M - v_m < 0$	$F_g = 0$
	$\dot{x}_M = v_M < 0$ $v_M - v_m > 0$	$F_g = 0$

In order to reproduce the sky-hook concept (the force generator delivers a force $F_g = -D\dot{x}_m$) by means of the force of the semiactive damper, the following condition must apply (subject to physically realizable application of forces):

$$F_g = f_\tau(u) = D\dot{x}_M - c|\dot{x}_M - \dot{x}_m| \tag{6.36}$$

Table 6.3 schematically shows the actuation conditions to implement a semiactive sky-hook vibration control scheme.

Table 6.4 Conditions of activation in a semiactive relative control scheme.

Equivalent diagram	Condition	Damper force, F_g
	$\dot{x}_M - \dot{x}_m > 0$ $v_M - v_m > 0$	$F_g = 0$
	$\dot{x}_M - \dot{x}_m < 0$ $v_M - v_m < 0$	$F_g = 0$
	$\dot{x}_M - \dot{x}_m > 0$ $v_M - v_m < 0$	$F_g = D_r(\dot{x}_M - \dot{x}_m)$
	$\dot{x}_M - \dot{x}_m < 0$ $v_M - v_m > 0$	$F_g = D_r(\dot{x}_M - \dot{x}_m)$

6.3.2 Relative vibration isolation

In the relative control approach, the condition of actuation by means of the dissipative force in the semiactive damper relates not only to the relative velocity but also to the state of the spring. The relative control law is formulated thus:

$$f_{damper} = D_r \cdot (\dot{x}_M - \dot{x}_m) \quad \text{if} \quad (x_M - x_m) \cdot (\dot{x}_M - \dot{x}_m) \leq 0$$

$$f_{damper} = 0 \qquad\qquad \text{if} \quad (x_M - x_m) \cdot (\dot{x}_M - \dot{x}_m) > 0 \qquad (6.37)$$

For practical purposes, the relative control approach requires at least one position sensor to measure the relative position of the damper. For its part, the sky-hook control approach requires determination of the absolute velocity of the sprung mass, which is not always readily available. Table 6.4 schematically shows the actuation conditions to implement a semiactive relative vibration control scheme.

6.4 Figures of merit of ER and MR devices

For practical application of ER and MR devices, the designer needs an indication of the relative performance of both technologies. The relevant set of design criteria includes the operating temperature range, the maximum dissipative force (or pressure drop), the size and weight of the actuator, its power requirements, the power density and the driving conditions. These figures of merit are analyzed in detail in the following paragraphs.

6.4.1 Material aspects

Some of the material characteristics of ER and MR fluids were introduced in earlier sections in the context of the comparative analysis of field-responsive fluids (thus including ferrofluids).

The basic viscosity, η, is similar for both ER and MR fluids, of the order of 0.2–0.3 Pa·s. The maximum yield stress, τ_y, is of the order of 2–5 kPa for ER fluids and 50–100 kPa for MR fluids.

In the case of ER fluids, the maximum applicable electric field across the active fluid is around 3–5 kV/mm; this is limited in practice by the dielectric breakdown electric field (which produces electrical current flow across the fluid). In addition, the conductivity of the fluid doubles with every temperature increase of 6 °C. The maximum applicable magnetic field for MR fluids is of the order of 150–250 kA/m. In this case, the practical limitation comes from the magnetic saturation of the fluid.

In addition to these practical limitations, ER fluids are extremely sensitive to external contaminants and temperature conditions.

6.4.2   Size and weight of ER and MR devices

Here, we will recall Equation 6.20, which established the relationship between the required active volume of ER–MR fluids and the application requirements in terms of mechanical power to be dissipated and mechanical control ratio:

$$V = K \frac{\eta}{\tau_y^2} \lambda P_m$$

If we consider a particular application, for example, a set of requirements in terms of mechanical power, P_m, and control ratio, λ, the relationship between the active volume for the ER, V_{ER}, and the MR solution, V_{MR}, is

$$\frac{V_{ER}}{V_{MR}} = \frac{\left(\frac{\eta_{ER}}{\tau_{ER}^2} \right)}{\left(\frac{\eta_{MR}}{\tau_{MR}^2} \right)} \qquad (6.38)$$

Since the basic viscosity for ER and MR fluids is similar, $\eta_{ER} \approx \eta_{MR}$, Equation 6.38 can be rewritten as follows:

$$\frac{V_{ER}}{V_{MR}} \approx \left(\frac{\tau_{MR}}{\tau_{ER}} \right)^2 \qquad (6.39)$$

If we consider that yield stress is 1 order of magnitude larger in MR fluids than in ER fluids, the requirements in terms of active volume are between 100 and 1000 times higher for ER than for MR fluids.

The above analysis considers only the volume of the active fluid. ER and MR devices will include additional elements to set up the electric and magnetic fields respectively. These external components are slightly larger in the case of MR devices (mainly due to the size of the coils used to establish the magnetic field). However, the qualitative analysis is still valid: because the yield stress is higher in MR fluids than in their ER counterparts, the size of the former can be considerably less. Moreover, this has implications for the scaling properties of ER and MR devices.

The density of MR fluids is typically twice that of ER fluids. However, if the required active volume is considered, the overall weight of the MR device will be lower than the weight of the ER device.

6.4.3   Available dissipative force and power

In this section, we return again to Equation 6.20 to establish a comparison between ER and MR devices in terms of dissipative force and power. Equation 6.20 can be rearranged to provide an expression for the dissipation power density, $P_V = P_m / V$, of ER and MR fluids:

$$P_V = \frac{\tau_y^2}{K \lambda \eta} \qquad (6.40)$$

The dissipative energy density per cycle, W_V, can be derived from Equation 6.40 by considering that the power density is the product of the energy density multiplied by the frequency of operation, $P_V = W_V \cdot f$. Since the maximum frequency of operation is similar for ER and MR devices, we end up with the following result:

$$\frac{P_{V_{ER}}}{P_{V_{MR}}} = \left(\frac{\tau_{MR}}{\tau_{ER}}\right)^2 \tag{6.41}$$

This result once again demonstrates the superiority of the behavior of MR fluid–based dissipative devices over that of EM-based ones. A similar analysis can be applied to the available dissipative force, which is related linearly to the maximum yield stress.

A direct comparison of the electrical energy required to drive ER and MR devices is difficult. Electrical energy losses in ER devices depend on current drawn through the fluid, while in the case of MR devices the electrical loss is caused by Joule dissipation at the coils.

A comparison is possible on the basis of the energy required to set up the electric field (charge the capacitor) in ER devices and the magnetic field in MR devices. The electrical energy required for ER devices can be computed from the product of active volume, V_{ER}, and energy density in the control volume:

$$W_{ER} = V_{ER} \left(\tfrac{1}{2}\kappa\epsilon_0 E^2\right) \tag{6.42}$$

where κ is the relative dielectric coefficient for the ER fluid and ϵ_0 is the vacuum permittivity.

Similarly, the energy required to establish the magnetic field across the active volume, V_{MR}, in the MR device is

$$W_{MR} = V_{MR} \left(\tfrac{1}{2}BH\right) \tag{6.43}$$

According to Equations 6.42 and 6.43, the ratio of energy required in ER devices to energy required in MR devices is

$$\frac{W_{ER}}{W_{MR}} = \frac{V_{ER}}{V_{MR}} \left(\frac{\kappa\epsilon_0 E^2}{BH}\right)^2 \tag{6.44}$$

If a volume ratio of approximately 100 is assumed, then taking into account that typical values for all the other parameters are $\kappa = 10$, $\epsilon_0 = 8.85 \times 10^{-12}$ F/m, $E = 4$ MV/m, $B = 0.8$ T and $H = 2 \times 10^5$ A/m, the energy ratio is as follows:

$$\frac{W_{ER}}{W_{MR}} \approx 1 \tag{6.45}$$

The practical interpretation of this result is that, for the same required dissipative power, the energy input required by ER and MR devices is approximately the same. The main difference is that while ER devices require high-voltage sources, MR devices are driven with readily available low-voltage, high-current power sources.

6.4.4 Scaling of active rheology concepts

In this section, we analyze the scaling properties of ER and MR fluid actuators. The analysis focuses on force and stroke, dissipative energy, dissipative power and time constant or frequency bandwidth. Finally, there are some comments regarding the size of the particles in the fluids as a factor limiting miniaturization.

Our analysis considers all three basic configurations (flow, shear and squeeze mode devices). It is shown that the three configurations present the same scaling properties.

Force and stroke upon scaling

The different expressions for the viscous and the field-dependent force and pressure drop were given in Section 6.2.1 (see Equations 6.14 to 6.19). The scaling analysis for the forces in the shear and squeeze modes yields the following expressions:

- *Shear Mode.* From Equations 6.14–6.15, the following scaling laws can be found:

$$F_\eta = \frac{\eta \dot{x} b a}{g} \propto L \tag{6.46}$$

and

$$F_\tau(E, H) = \tau_y(E, H) b a \propto L^2 \tag{6.47}$$

The viscous force F_η (which is the minimum force the device will apply with no external field) scales as L, while the field-dependent force, F_τ, scales proportionally to L^2.

The control ratio, λ, which was defined as the ratio of F_τ to F_η, will therefore scale as $\lambda \propto L$.

The volumetric forces will scale down according to the following expressions:

$$\{F_\eta\}_V \propto L^{-2} \tag{6.48}$$

$$\{F_\tau(E, H)\}_V \propto L^{-1} \tag{6.49}$$

A glance at the above expressions will show that upon miniaturization the viscous forces (which are to be kept to a minimum) scale down less well than the field-dependent forces. Thus, the control ratio is reduced as the device is miniaturized, and consequently the 'ON' state forces become ever less significant as the device scales down.

- *Squeeze Mode.* Applying a similar analysis as in the previous case, we find the following set of expressions:

$$F_\eta = \frac{3\pi \eta R^4 \dot{x}}{2(g_0 - x)^3} \propto L \tag{6.50}$$

and

$$F_\tau(E, H) = \frac{4\pi \tau_y(E, H)R^3}{3(g_0 - x)} \propto L^2 \qquad (6.51)$$

and the volumetric forces reduce to the following scaling laws:

$$\{F_\eta\}_V \propto L^{-2} \qquad (6.52)$$

and

$$\{F_\tau(E, H)\}_V \propto L^{-1} \qquad (6.53)$$

which are the same scaling laws as in the shear mode analysis. Consequently, the same conclusions can be applied to the squeeze mode concept.

In the case of *flow mode*, Equations 6.16 and 6.17 were developed to express the relationship between the flow across the actuator's active area and the pressure drop. For a coherent scaling analysis, these equations must be translated into expressions relating device force to velocity. To do so, we note that the force is simply the pressure drop times the cross- sectional area ($A \propto L^2$), and that the volumetric flow is the cross-sectional area times the device's velocity. Hence, $Q \propto L^2$:

$$F_\eta = A \cdot P_\eta(S) \propto L^2 \qquad (6.54)$$

and

$$F_\tau = A \cdot P_\tau(E, H) \propto L \qquad (6.55)$$

which is again the same scaling law as in the previous two modes. The volumetric forces scale according to the following expressions:

$$\{F_\eta\}_V \propto L^{-2} \qquad (6.56)$$

and

$$\{F_\tau(E, H)\}_V \propto L^{-1} \qquad (6.57)$$

The scaling law for the stroke is the same for the three design configurations and produces a linear relationship between the device's stroke and its dimensions, $S \propto L$.

Energy and power density upon scaling

Let us start the derivation of scaling laws for the energy density and the power density by recalling expression 6.20, which stated the active volume requirements for a given power dissipation:

$$V = K \frac{\eta}{\tau_y^2} \lambda P_m$$

where P_m is the mechanical power to be dissipated and V is the active volume required for the actuator. The appropriate rearrangement of this equation gives us

$$P_V = \frac{P_m}{V} = \frac{\tau_y^2}{K\eta}\lambda^{-1} \tag{6.58}$$

In the previous sections, we saw that the control ratio for all ERF and MRF actuator configuration scales as $\lambda \propto L$, and, therefore, the scaling law for the power density is

$$P_V \propto \lambda^{-1} \propto L^{-1} \tag{6.59}$$

This expression indicates that the volumetric density of the power dissipated by the actuator increases linearly as the actuator dimensions are reduced.

The scaling law for the volumetric energy density, W_V, can be arrived at by considering that the energy is the dissipative force times the displacement. Since scaling laws have already been developed for force and stroke in the previous sections, the operation is quite straightforward:

$$W_V = \frac{F \cdot S}{V} \propto L^0 \tag{6.60}$$

Time constant and frequency bandwidth upon scaling

Since the power dissipated and the energy or work dissipated per cycle are related through the frequency, the scaling law for the frequency of actuation with ERF and MRF fluids will be as follows:

$$f = \frac{P_V}{W_V} \propto L^{-1} \tag{6.61}$$

This indicates that miniaturized devices are expected to be faster. Given that the frequency and the actuator time constant are inversely proportional, from Equation 6.61 it follows that the scaling law for the actuator's time constant is

$$\tau \propto \frac{1}{f} \propto L^1 \tag{6.62}$$

Other scaling considerations

Section 6.4.2 offered some general remarks on the relative size of ERF and MRF actuators. No assumption was made about the type of responsive fluid in the above derivation of scaling laws. These are therefore general and applicable to both ERF and MRF actuators.

In the development of scaling laws, other material-related aspects may play a crucial role. For example, in ERF and MRF actuators, the size of the particles will impose a practical limit to miniaturization. In particular, as explained in Section 6.1.2 (see Table 6.1), the particles in ER fluids are 1 order of magnitude larger than those

Table 6.5 Operational characteristics and scaling trends for ERF and MRF actuators.

Figures of merit	MRF actuators	ERF actuators
Force, F	$10^2 \leq 10^5$ N	100 times lower
Displacement, S		$10^0 \leq 10^3$ mm
Work density, W_V		$10^{-3} \leq 10^{-2}$ J/cm^3
Power density, P_V		$10^{-2} \leq 10^0$ W/cm^3
Bandwidth, f		Similar for ERF and MRF
Size		Active volume 10^2–10^3 times lower for MRFs
Energy consumption		Up to 10^3 Hz
Scaling trends		
Force		$F_\eta \propto L$ and $F_\tau \propto L^2$
Stroke		$S \propto L$
Work per cycle		$W \propto L^3$
Energy density		$W_V \propto L^0$
Bandwidth		$f \propto L^{-1}$
Power density		$P_V \propto L^{-1}$

in MR fluid. In general, this limits the minimum gap size for all three actuator configurations. Therefore, the possibilities of miniaturization will, in principle, be less restricted in MR fluids than in ER fluids. Table 6.5 shows a summary of the most important figures of merit and scaling laws for ERF and MRF actuators.

6.5 Applications

Case Study 6.1: Magnetorheologic lower limb prosthesis for improved gait performance

Human gait is a highly dynamic, specialized operation in which a set of functional actions are cycled. Human gait is usually studied as a cycle in which two main phases can be distinguished, namely, stance and swing. During stance, the lower limb musculoskeletal system supports the body weight, absorbs the shock produced by foot contact and provides the torque required to extend the knee prior to the next phase. During swing, the knee is initially flexed to stop the foot from contacting the ground and, finally, fully extended in preparation for the next step.

The relationship between the kinetic and the kinematic characteristics of the knee articulation depends on the walking cadence, walking conditions (rough terrain, slopes, stairs, soft floors, etc.) and turns. Ideally, prosthetic solutions for above-knee amputees should mimic natural walking conditions as closely as possible.

As an example of a state-of-the-art active prosthesis, the Otto Bock C-Leg®
prosthetic knee includes multiple sensors. These transmit information at a speed of
50 Hz, allowing the feedback controller to operate its mechanical and hydraulic systems. Two strain gauges measure pressures on the leg and determine how often the
heel strikes (thus giving an estimation of the walking cadence); magnetic sensors
report changes in knee angle.

The use of field-responsive fluids in orthotic and prosthetic applications is a
logical step in replacing traditional hydraulic or pure electromechanical systems.
One recent introduction is the Prolite™ Smart Magnetix™ Above-the-Knee (AK)
Prosthesis. Prolite™ Smart Magnetix™ is manufactured by Biedermann OT Ver-
trieb, a German manufacturer of prosthetic components, and was developed jointly
with Lord Corporation. Part of the information in this section was provided by
Lord Corporation. The pictures in this section are courtesy of G. Hummel and
L. Yanyo, Lord Corporation.

Figure 6.15 shows a schematic representation of the Prolite™ Smart Magnetix™
system. The system includes both kinetic (force and torque) and kinematic (angular
position and rate) sensors. The sensors are used to adapt the rheological characteristics of a modified Lord RD-1005 MR fluid damper (technology adapted from
Lord's Motion Master® Ride Management System truck seat damper, see Case
Study 6.3).

The system incorporates controllers that adapt the damping characteristics of
the MR damper to the walking conditions. Owing to the fast response time of the

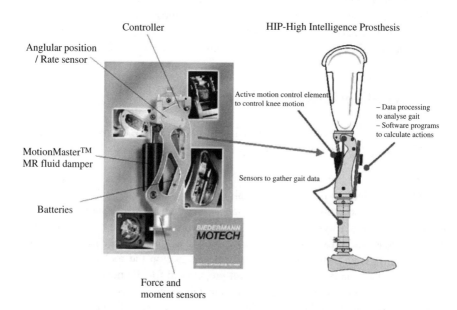

Figure 6.15   Schematic representation of the Prolite™ Smart Magnetix™ Above-
the-Knee (AK) Prosthesis (Reproduced by permission of Lord Corporation).

Figure 6.16 A Prolite™ Smart Magnetix™ Above-the-Knee (AK) Prosthesis user walking down a slope: the characteristics required for the knee prosthesis will generally be different from other walking conditions. The MR technology provides the desired adaptation (Reproduced by permission of Lord Corporation).

MR technology, this adaptation can be made very quickly (at a rate of 500 Hz), which results in a more natural gait and makes climbing up and down stairs and slopes much easier. Moreover, this makes for a more efficient walking pattern, which is one of the most serious problems suffered by users of passive prostheses.

Figure 6.16 shows a prosthesis user walking down a slope with the Prolite™ Smart Magnetix™ fitted. It is claimed that after the adaptation process, users can ride a bicycle, carry heavy objects and walk or run with varying gaits.

Case Study 6.2: Adjustable orthotic suppression of pathologic tremor

Tremor is a symptom indicating the possibility of a neurological disorder. Tremor is regarded as a movement disorder that can affect all body parts, but it is most common in the upper limbs, trunk and head. Tremor of itself is highly incapacitating, sometimes making it practically impossible for the patient to perform activities of daily living (ADL).

There are several treatments for tremor, but amongst the most common are medication, rehabilitation programmes and deep brain stimulation, DBS (which is an invasive technique involving surgery to implant microelectrodes in appropriate structures in the brain). Alternatively, it has been recognized that the modification of inertial (mass), damping and stiffness biomechanical characteristics of the upper limb alters the tremor pattern. Appropriate selection of these parameters may even reduce tremor to acceptable levels.

This section briefly describes the application of MR technology to orthotic suppression of tremor. The European Commission has recently funded research into alternative orthotic solutions for reducing tremor through the project *DRIFTS, Dynamically Responsive Intervention for Tremor Suppression* under contract *QLK6-CT-2002-00536*. The project (of which the author is an active member) is coordinated by L. Normie (GeronTech–The Israeli Centre for Assistive Technology and Ageing), and the work on the application of MR technology to counteract tremor is led by W. Harwing (Department of Cybernetics, University of Reading).

Pathological tremor of the upper limbs can be classified into three categories according to the type of movements that are affected, that is, rest tremor (typical of Parkinson's disease), postural tremor (typical of patients with essential tremor) and kinetic tremor.

Rest tremor produces rhythmic oscillation of the arm in a frequency range between 3 and 6 Hz and is only present when the limb is resting. Postural tremor is characterized by oscillations in a range from 5 to 12 Hz and is evident when the arms sustain a posture (for instance, outstretched hands). Finally, kinetic tremor occurs in a frequency range between 2 and 4 Hz and is associated with intentional upper limb motion.

The DRIFTS concept is based on the application of a shunt load to the articulations of the upper limb affected by tremor. Previous studies indicate that additional damping of the limb reduces the tremor. Since the characteristics of each patient's tremor are different, a general orthotic device for cancelling tremor should have the ability to adapt the damping parameters to the particular tremor.

Several alternatives have been analyzed within the framework of DRIFTS. In particular, MR fluids were chosen from amongst various different actuator technologies as the basis of the adaptable damper. Ultrasonic motors and electroactive polymers have also been evaluated as an interesting alternative for direct drive operation.

The wearable tremor suppression orthosis consists in mechanical transmission linking the intrinsically rotational movement about an articulation (the wrist in first instance) to a linear adjustable MRF damper (see Figure 6.17). The damper itself is secured to the dorsal part of the forearm by means of adjustable textiles (see Figure 6.18).

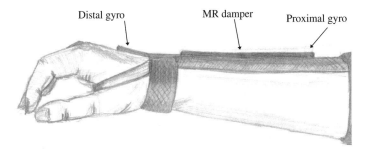

Distal gyro   MR damper   Proximal gyro

Figure 6.17   Schematic representation of the DRIFTS MRF damper structure.

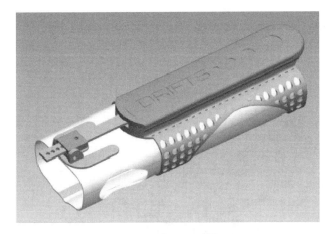

Figure 6.18   CAD model of the wrist tremor suppression orthosis.

The MRF damper is based on the shear mode concept described in previous sections. A printed coil circuit in the upper part of the shear plate generates the required magnetic field over the active region of the MR fluid (see the schematic view in Figure 6.19).

The MRF damper has been designed to deliver a maximum damping force of around 30 N. This is transmitted to the wrist, where a maximum dissipative torque of 1 Nm is achieved. The magnetic circuit is so designed that a continuous electrical current of up to 1 A (peak currents of up to 2 A) can be applied to the coil.

In the development of the tremor cancellation MR orthosis, available and specifically developed MR fluids and damper configurations were analyzed thoroughly. The system is equipped with gyroscopes at both the proximal and the distal parts of the device. These sensors permit differential measurement of the angle, velocity and acceleration of wrist rotation. The system further includes force sensors, so

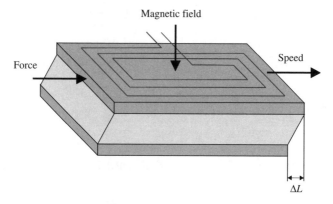

Figure 6.19   Damper scheme and electric coil to set up the magnetic field.

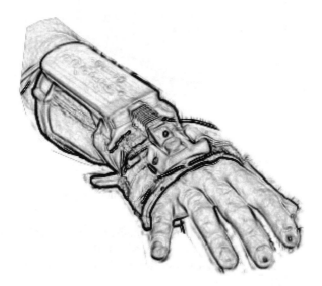

Figure 6.20   MR fluid–based tremor reduction orthosis.

that it can be programmed to mimic a passive adjustable viscosity damper (see Figure 6.20).

Case Study 6.3:   Semiactive damping of the driver's seat in heavy goods vehicles

Ergonomics at the workplace is a common concern today. Truck drivers in transport and logistic companies are subject to continuous stress because of shock and vibrations transmitted to their seats from road-induced shock and vibrations. In

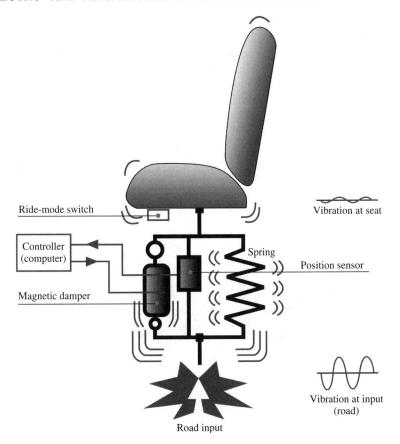

Figure 6.21   Schematic representation of the Motion Master® Ride Management System: shock and vibration transmitted from the road are monitored by the position sensor. This provides a set point for the MR-based damper force. The result is reduced seat vibration and hence the elimination of topping and bottoming (Reproduced by permission of Lord Corporation).

addition to possible injury, excessive vibration can compromise driving ability and lead to loss of control.

Comfort in trucks, and in heavy transport systems in general, relies on pneumatic and hydraulic suspended seats. Traditional suspension systems cannot adapt their damping characteristics to road conditions and to the driver's weight.

Lord Corporation recently introduced the Motion Master® Ride Management System. Motion Master® is based on magnetorheological fluid technology and provides a semiactive damping system for improved driver's seat comfort and safety. The figures in this section are courtesy of G. Hummel, Lord Corporation. Part of the information used in this case study was likewise provided by Lord.

Figure 6.21 is a schematic representation of the operating principle of the Motion Master® system. Like any other seat suspension system, it includes a pneumatic base spring and a damper. In this case, the traditional passive damper is replaced by a magnetorheological damper. The system further incorporates a position sensor, a controller and a user switch.

In the most common traditional situation, the seat suspension in trucks is tuned to absorb vibrations in a frequency range of 4 to 8 Hz, which has proven to be the most potentially harmful range for drivers. Traditional systems cannot deal effectively with bad road conditions and usually produce topping and bottoming shocks.

The concept of a MR-based seat suspension was depicted in Figure 6.21. The traditional passive damper is replaced by the Lord RD-1005 MR fluid damper. A position sensor is adapted to the seat suspension system to monitor the variable

Figure 6.22   Constituent parts of the Motion Master® system: position sensor, active MR damper, controller and user switch (Reproduced by permission of Lord Corporation).

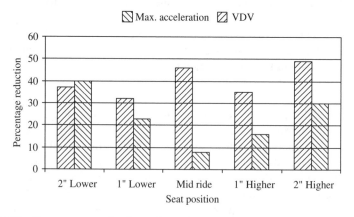

Figure 6.23   Percentage reduction of maximum acceleration and vibration dose value (VDV) as a function of seat position (Reproduced by permission of Lord Corporation).

vibration level. This information is used in a sky-hook control strategy to adjust the damping force. In addition, the system incorporates a user switch that allows the driver to set the damping characteristics of the system to firm, medium or soft conditions. The various different components in the Motion Master® semiactive shock and vibration absorption system are shown in Figure 6.22.

The vibration absorption results, as reported by Lord Corporation, indicate that the reduction in the maximum acceleration transmitted to the driver is as high as 49%. Figure 6.23 shows the percentage reduction in acceleration as a function of the seat position. Motion Master® was introduced in 1998 and is currently being used in thousands of vehicles. It may therefore be regarded as one of the successes of MR technology application.

7

Summary, conclusions and outlook

This chapter summarizes the most important features of all the emerging actuator technologies discussed in the book and subjects them to a comparative analysis. Here, traditional actuators are included as a reference for comparative purposes. The author concludes that emerging and traditional actuators should be considered as complementary rather than competing technologies.

The last part of the chapter analyzes the probable evolution of the different technologies, highlighting open research topics and indicating application domains.

7.1 Brief summary

This book began with an introduction to *actuators as a special case of transducing devices* in which energy is converted from the electrical to the mechanical domain. As such, they are usually incorporated in motion control systems, where they link the control action to the mechanical action on the plant.

A direct consequence of the linking position of the actuator between control and plant is that *both electrical and mechanical impedance matching are required.* This was exemplified in Chapter 1, which examined each of the particular emerging actuator technologies discussed in the book.

The book makes a clear distinction between *active actuators* (described in detail in Chapters 2 through 5) and *semiactive* actuators (described in Chapter 6). Active actuators are those whose action on the plant may cause energy to be added or subtracted. Semiactive actuators, on the other hand, can only dissipate (remove) energy from the plant.

Actuators have been *presented as mechatronic systems.* Emerging actuators are reported to combine both sensing and actuation functions. Given their shared

Emerging Actuator Technologies: A Micromechatronic Approach   J. L. Pons
© 2005 John Wiley & Sons, Ltd

functionality, they meet the definition of true mechatronic systems, and as such all the principles of mechatronic and concurrent engineering apply to them.

The applicability of the mechatronic concept to actuators leads naturally to the notion of *smart actuators as reactive devices* that are able to autonomously optimize their operating point in response to external or internal disturbances. This was exemplified, in particular, in Chapters 2, 3 and 5 when dealing with piezoelectric, SMA and magnetostrictive actuators respectively.

The *underlying biomimetic approach to the design and control of emerging actuators* was stressed in Chapter 1, where nature was identified as a source of inspiration in the context of the following:

- Design of actuation concepts, in particular, in inchworm and travelling wave piezoelectric actuators.

- Switched control of some nonlinear, varying-dynamics actuators, in particular, SMAs and EAPs.

- Hierarchical organization of control architecture mimicking the motor control hierarchy in humans.

- Responsive smart structures and actuators mimicking the reflex control of human limbs.

Emerging actuators have been defined as novel actuator technologies based on transducing materials. The advent of new transducing phenomena may eventually lead to the formulation of new actuator technologies. However, the process may take a long time.

For a new transducing phenomenon to be brought up the stage of practical implementation and eventually to feasible actuators, the following is essential:

- *To Master the New Phenomenon.* To identify all the implications of the new transduction until constitutive equations that fully describe the phenomenon are finally arrived at.

- *To Master the Material Aspects.* The new transduction level (energy out to energy in) may not be appropriate with existing materials. New materials offering enhanced transduction must be engineered, for instance, in the case of the magnetostriction and the giant magnetostrictive effect in Terfenol-D, see Chapter 5.

- *To Develop Appropriate Actuation Concepts.* Even when the transduction phenomenon and the materials are there, actuation concepts may not be apparent. This is true, for instance, of travelling wave ultrasonic motors, which have only recently been developed out of piezoelectric materials.

Figure 7.1 illustrates the evolution of the various different emerging actuator technologies described in this book. Some of them (see, for example, the case of

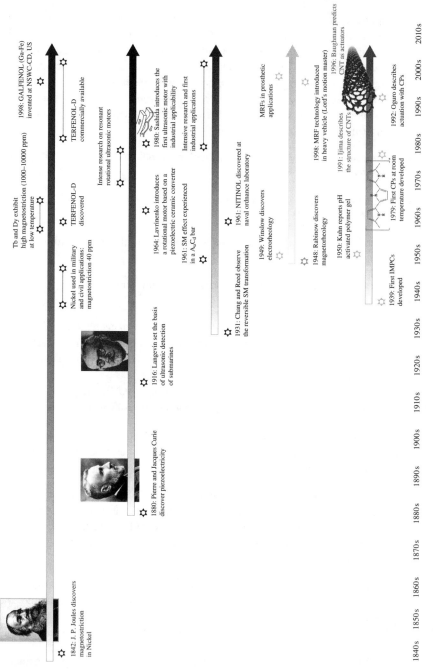

Figure 7.1   Chronological evolution of emerging actuators: from transducing materials to actuator development.

piezoelectricity and magnetostriction) have evolved over the last 150 years until, finally, appropriate actuator technologies have begun to emerge.

Throughout this book, traditional actuator technologies have been used as a reference for the comparative analysis of emerging actuators. The following sections are intended to illustrate the comparative positions of the various emerging actuator technologies, and there again traditional actuators (electromagnetic, pneumatic and hydraulic) are cited as references. The *key concept that this comparison elicits is that of complementarity.* Traditional and emerging actuators should be seen not as competing but as complementing one another, so that the advent of new technologies opens up new application domains.

The next few paragraphs summarize the most outstanding features of each emerging actuator technology.

7.1.1 Piezoelectric actuators

Emerging actuators based on piezoelectric transduction are *the best established of all the emerging technologies.* They are based on materials exhibiting piezoelectricity (a phenomenon known since the nineteenth century).

There are manifold implementations of piezoelectric actuators. They can be *classified into resonant and nonresonant piezoelectric actuators.* The technology spans a wide range of stroke, force, energy, power and time response levels, depending on the particular implementation.

Resonant piezoelectric actuators are *based on the frictional transmission of microscopic elliptic vibrations (excited in resonance) between a stator and a rotor.* They can be developed in both linear and rotational versions. Usually the stroke is large and the force, energy and power density are moderate.

Nonresonant piezoelectric motors use either the piezoelectric phenomenon directly to drive the load, or a geometric concept to transform the piezoelectric effect into a macroscopic motion. In either case, they are driven out of resonance.

The main *control issues* in resonant piezoelectric actuators are maintaining the resonant operating point irrespective of disturbances and matching the electrical impedance between the controller and the actuator for efficient operation. The crucial control aspects in nonresonant drives are compensation of hysteresis and linearization.

The smart actuator concept can be readily developed with piezoelectric materials since piezoelectricity is a two-way transducing phenomenon.

The properties of this actuator technology and its evolution upon scaling were discussed in Chapter 2. They are studied comparatively in Section 7.2.

7.1.2 Shape memory alloy actuators

Shape memory alloy actuators are thermally triggered devices and are therefore *slow drives.* They are classed as *half-cycle, soft actuators* because they can only provide pulling operation. The other half-cycle is completed by a load-driven deformation of the alloy.

SMAs exhibit other interesting effects in addition to the shape recovery phenomenon. They *exhibit pseudoelasticity (also called superelasticity)*, a dissipative hysteretic phenomenon whereby high strains (up to 10–15%) can be recovered (like elasticity but on a much larger scale). These alloys also undergo a fivefold change in Young's modulus during the thermal transformation.

Actuator concepts can be developed from *different load configurations* (tensile, bending, torsion and combinations), so that there is a large spectrum of strokes and forces. In the design process, the thermal and material configurations can be optimized for faster response and suitability to the application.

The most *important control issues* in dealing with SMA actuators are their extreme nonlinearity and hysteretic behavior, and concurrent use of this technology to develop the sensor function. The latter is usually based on electrical resistance to strain models for the actuator.

SMA actuators are high-force drives (having the highest force density of all technologies). They are slow actuators whose properties improve when their dimensions are scaled down. The lifetime for this technology is severely limited by material fatigue. Because of the combination of these actuation properties, they are more suitable for one-time applications or passive thermal actuators triggered by changes in ambient temperature.

7.1.3 Electroactive polymer actuators

Electroactive polymer actuators are the newest of all the driving technologies described in this book. There are no commercial applications of any of the different implementations of this technology, and it can therefore be considered as being still in the preliminary stages.

The category of EAP actuators is an umbrella term covering a large number of transducing materials and principles. All of them still require further study for the actuation mechanisms to be fully understood. Too many material and design unknowns are still the subjects of scientific controversy for them to be described as truly emergent at this time.

EAP actuators can be classified into wet (ionic) and dry (electronic) polymer actuators. Wet actuators are slow (limited by transport mechanisms) and are excited by low voltages. Dry actuators are faster but require very intense electric fields.

Control of EAPs varies considerably from one technology to another. In most cases, polymer gels exhibit discontinuous phase transitions that make it impossible to control position in equilibrium, and, therefore, switching techniques are required. Ionic polymer–metal composite (IPMC) actuators present varying dynamics upon contraction and subsequent relaxation and are therefore also better controlled by switching techniques. Carbon nanotube (CNT) actuators, conducting polymer (CP) actuators and the dry EAP technologies do not require significantly complex control strategies.

It is the author's opinion that suitable application niches should be sought for the various different EAP actuators. These (especially wet EAPs) do not seem to be suitable at this stage to replace or compete with traditional or other emerging actuator technologies. Rather, their use should be restricted to environments where their particular drive actuation conditions are readily available – for instance, CP actuators in biomedical applications.

The actuation characteristics can only be given at a theoretical level. The characteristics in the comparative charts in Section 7.2 are only theoretical, upper-limit estimations. Real figures will most likely be several orders of magnitude lower than these theoretical expectations.

7.1.4   Magnetostrictive actuators

Magnetostriction is a good example of a long time lapse between the discovery of a phenomenon (transduction, 1841) and the engineering of appropriate materials (giant magnetostriction in the 1960s).

With its small to moderate stroke (up to 3000–4000 ppm in dynamic driving), high force and high bandwidth, this technology is best suited to vibration control applications. The technology and basics of active vibration control were described in detail in Chapter 5.

Like many of the emerging actuator technologies, magnetostrictive actuators are also suitable for use as sensors. They are particularly well adapted for integration in truss-type smart 3D structures. Chapter 5 also devotes a complete section to the actuation principles of this and other technologies in smart structures.

In addition to the actuation principle (i.e. magnetically triggered generation of strain), magnetostrictive materials undergo a magnetically induced change in stiffness, and, consequently, in resonance frequency. These effects are useful for the development of adaptable structures and tunable vibration absorbers.

7.1.5   Electro- and Magnetorheological fluid actuators

Electro- and magnetorheological fluid actuators are the only semiactive actuators described in this book. They are based on field-responsive fluids in which the rheological properties are modified as a result of an applied electric or magnetic field.

There are three possible concepts for the development of ERF and MRF actuators: shear, flow and squeeze mode devices. For a proper approach to the design, control and application of these actuators, an extended mechatronic definition is required that includes the magnetic domain.

In such actuators, the operation point is typically chosen close to the magnetic saturation of the fluid (thus fully exploiting the characteristics of the materials). A reluctance circuit is designed and used to focus the magnetic flux into the active volume. The equivalent electrical load of the actuator is inductive (for MRF) or capacitive (for ERF), depending on the type of fluid, and so the appropriate design criteria have to be developed.

Being semiactive, these actuators can only remove energy from the plant and so they are used in semiactive vibration control applications. A full section of Chapter 6 is therefore devoted to the basic concepts of semiactive vibration control and two classic implementations of this control approach: sky-hook and relative control.

ERF and MRF actuators require approximately the same amount of input electrical energy to dissipate mechanical energy. However, because the transduction is greater in MRFs, their active volumes are 100–1000 times lower than those of ERFs. Given this and the fact that the suspended particles in the ERF fluid are larger, MRFs are the better suited of the two for miniaturization.

7.1.6   Example applications: case studies

This book is intended to bring together the theoretical background to the mechatronic analysis, design and control of emerging actuators and the practical aspects of their application. Therefore, each emerging actuator technology is illustrated with a set of case studies.

Wherever possible, case studies have been selected from among industrial and commercial applications. This ensures that the technology complies with the relevant set of industrial standards for the particular application. In some cases, research implementations are presented in addition to industrial applications. These examples were selected either because they exemplify an important aspect of the actuator not present in the industrial examples or simply because no other industrial applications were available. In the case of EAP actuators, all the selected applications are at the research stage.

Piezoelectric actuators

Five case studies are presented to exemplify the application of piezoelectric actuators. Four of these (Case Studies 2.1, 2.2, 2.4 and 2.5) are industrial implementations in different industrial domains. Case Study 2.3 is an implementation that is still at the research stage.

Case Studies 2.1 and 2.2 concern resonant piezoelectric actuators. The first describes a set of commercial OEM rotational ultrasonic motors with high torque and low-speed characteristics. The second is an application of the travelling wave ultrasonic motor principle to the optical auto-focus of Canon reflex cameras.

Case Studies 2.3, 2.4 and 2.5, on the other hand, describe different implementations of nonresonant piezoelectric drives: piezoelectric stepping actuators (based on piezoelectric stacks), piezoelectric benders and amplified piezoelectric stacks, respectively. These examples cover the industrial areas of optical appliances, machine tools, textiles and aerospace.

Shape memory alloy actuators

Four different case studies have been selected to exemplify the diverse application domains of SMA actuators. The first two examples, 3.1 and 3.2, are industrial implementations in the areas of aerospace and high-speed rail transport respectively. The last two Case Studies, 3.3 and 3.4, are research implementations.

Case Study 3.1 is an example of one-time actuation mechanism. It is an SMA-activated latching device for delicate instruments in aerospace applications. Once the instrument is in space, the actuator is used to release a latching device.

Case Study 3.2 is an example of thermal SMA actuators triggered passively by changes in the ambient temperature. It is used to keep the oil temperature in gearing systems of high-speed trains within acceptable limits.

Case Studies, 3.3 and 3.4 are included to illustrate the active use of SMA actuators and to introduce an application domain that is typical in the scientific literature on SMAs, namely, the biomedical application field. Case Study 3.3 is the first reported SMA-actuated endoscope and Case Study 3.4 is an implantable drug delivery device with SMA-triggered drug release.

Electroactive actuators

It was difficult to select applications for EAP actuator technology. No implementation of this novel technology has reached the market so far, and, therefore, the three case studies presented in Chapter 4 are all still at the research stage.

All three case studies concern the same EAP technology, as it is probably the best developed of all EAP actuator technologies: conducting polymer actuators. Case Study 4.1 describes the application of bilayer CP actuators in the biomedical field, specifically the case of microanastomosis of severed small blood vessels.

Case Studies 4.2 and 4.3 explain how linear and volume changes in CP actuators can be used to develop tactile displays and microvalves respectively.

Magnetostrictive actuators

As was comprehensively discussed in Chapter 5, magnetostrictive actuators are commonly found in active high-power vibration control applications. Therefore, Case Study 5.1 addresses the application of this technology to helicopter blade vibration control, as a classic example of its application.

Within the category of magnetostrictive actuators, a novel technology is evolving: magnetic shape memory actuators. These are characterized by a large stroke (of the order of 50,000 ppm) and moderate bandwidth at low force. We have included a short discussion of this new technology in Chapter 1. This discussion is supplemented by Case Study 5.2, which deals with the first prototype MSM actuators.

Electro- and Magnetorheological fluid actuators

The chapter devoted to ERF and MRF actuators includes three case studies. Case Studies 6.1 and 6.3 are examples of industrial implementations of MRF technologies in the prosthetic and automotive industry. Case Study 6.2 is an implementation in the orthotic domain, still at the research stage.

Case Study 6.1 is the only example not directly related to the field of vibration control. Instead, it presents the application of MRF actuators to optimal control of a knee prosthesis.

Case Studies 6.2 and 6.3 illustrate the typical application domain of ERF and MRF actuators: that is, vibration absorption. The latter introduces a high power dissipation application in heavy transport vehicle driver's seats, while the former describes the low-power application of MRFs to upper-limb orthotic tremor suppression.

7.2   Comparative position of emerging actuators

This section brings together the various emerging actuator technologies described in the book and compares their respective positions with one another and with traditional actuator technologies.

The data used to illustrate each feature were obtained from commercial catalogues where available and from scientific literature otherwise. For the case of EAP actuators, there is little information available as these are at a very early stage of development, and so theoretical projections are given instead of actual data.

7.2.1   Comparative analysis in terms of force

The maximum force is a figure of merit of actuator technologies that applies to piezoelectric (stacks, benders and inchworm type) actuators, shape memory alloy actuators, magnetostrictive actuators and electro- and magnetorheological actuators (in this last case, it is the maximum dissipative force).

Of the traditional actuator technologies, this figure of merit applies to linear electromagnetic motors and pneumatic actuators. Figure 7.2 shows a chart of the maximum available force versus the size of the actuator. It helps illustrate both the relative positions of emerging and traditional actuator technologies and the scaling trend.

The scaling analysis of force for all these technologies reveals a quadratic dependence on dimensions ($F \propto L^2$). The graph in Figure 7.2 includes lines showing this theoretical trend. There are only enough data points to indicate trends in the case of piezoelectric motors and traditional actuators, but these do match the theoretical predictions.

Piezoelectric stacks, SMA actuators and MS actuators can all be considered high-force drives ($F \geq 10^3$ N). Their force level is similar to that of pneumatic actuators.

Electromagnetic DC motors may be considered to generate moderate forces ($10 \leq F \leq 10^3$ N) and so effectively split the force spectrum into high-force and low-force actuators.

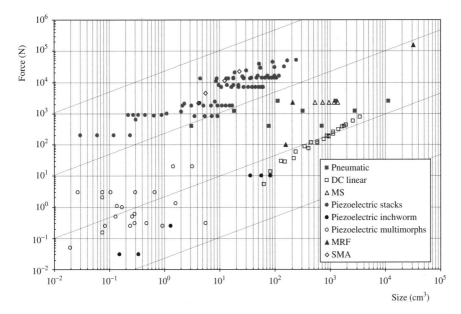

Figure 7.2   Comparison of emerging and traditional actuators in terms of force level.

Piezoelectric inchworm drives and benders may be considered representative examples of low-force actuators ($F \leq 10$ N). They generate virtually zero force and so are suitable for free positioning, precision devices.

7.2.2   Comparative analysis in terms of force density

Force density has been defined as the ratio of maximum available force to actuator volume or weight. This figure of merit is derived from the previous figure of merit and, thus, applies to the same type of actuators.

The chart in Figure 7.3 plots force density versus size for all the relevant emerging actuators and for linear electromagnetic motors and pneumatic actuators as reference for comparative purposes. Since the force density is plotted versus the actuator's size, it can be used to check scaling trends.

The theoretical scaling analysis for force density shows an inverse linear relationship with actuator dimensions ($F_V \propto L^{-1}$ N/cm^3); lines showing this theoretical trend have been plotted on the graph.

Again, SMA actuators and piezoelectric stacks may be considered high force density technologies ($F_V \geq 10^2$ N/cm^3). Magnetostrictive actuators belong in the medium force density range ($10 \leq F_V \leq 10^2$ N/cm^3) along with pneumatic actuators and piezoelectric multimorph actuators.

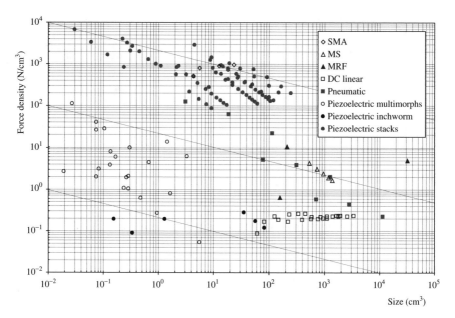

Figure 7.3   Comparison of emerging and traditional actuators in terms of force density level.

Electromagnetic DC linear motors and piezoelectric inchworm actuators exhibit the lowest force density ($F_V \leq 10$ N/cm^3).

7.2.3   Comparative analysis in terms of stroke

The stroke has been defined as the maximum available displacement for an actuator. This figure of merit applies to the same set of actuator technologies as the two previous cases. In emerging actuators, the stroke is typically determined as a fraction of the actuator's length. The following may be said about the stroke of emerging actuators:

- *Piezoelectric Stack Actuators.* These have the lowest stroke of all actuator technologies. In practice, it is limited to about 0.1–0.4% of the actuator's length for static applications. As Figure 7.4 shows, the stroke of these actuators is relatively low compared to any of the other technologies.

- *Piezoelectric Multimorph Actuators.* These have a higher stroke than stacked actuators or magnetostrictive actuators. Their stroke levels are moderate (see Figure 7.4).

- *Magnetostrictive Actuators.* The stroke for this technology is limited to about 1800 ppm for static applications. It is higher than the stroke of piezoelectric stacks but still low.

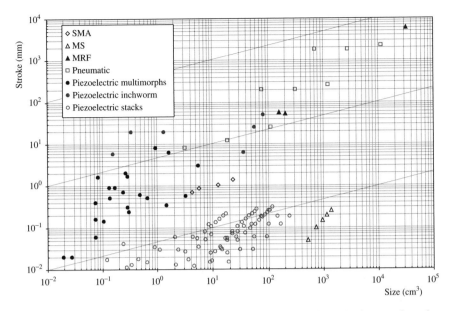

Figure 7.4   Comparison of emerging and traditional actuators in terms of stroke.

- *Shape Memory Alloy Actuators.* These and piezoelectric multimorph actuators present a moderate stroke, which, in practice, is limited to 3–4%.

- *Piezoelectric Inchworm Actuators.* These deliver the highest stroke of all piezoelectric technologies, similar to MRF actuators, and may be considered moderate to high.

- *Magnetorheological Fluid Actuators.* These span a large range of possible strokes, from moderate to high values.

- *Pneumatic Actuators.* These are shown here for comparative purposes. As Figure 7.4 shows, they have the highest stroke. DC electromagnetic linear motors are not shown since they can provide unlimited displacement if required.

Figure 7.4 shows a chart depicting the relationship between stroke and size in various different commercial implementations of all emerging actuator technologies. It can be used to analyze the scaling trend of stroke for these technologies (which is $S \propto L$). The lines describing the theoretical trend are included to assist such analysis.

7.2.4   Comparative analysis in terms of work density per cycle

Work density per cycle is one of the dynamic figures of merit described and analyzed throughout this book. It is the ratio of the maximum work per cycle to

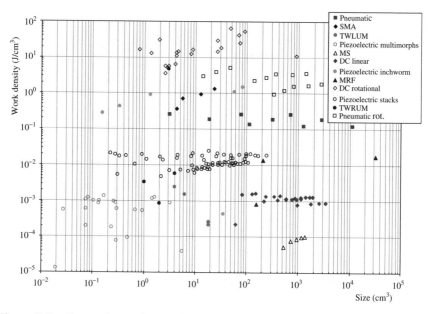

Figure 7.5 Comparison of emerging and traditional actuators in terms of work density per cycle.

the size or weight of the actuator. In the case of semiactive actuators, it represents the maximum energy dissipated per cycle.

The relative positions of the different actuator technologies for this figure of merit are represented graphically in Figure 7.5. The scaling trend for the work density per cycle indicates a roughly constant energy per cycle upon scaling ($W_V \propto L^0$). This feature can be readily verified by looking at the evolution of the various technologies for different actuator sizes.

Work density per cycle is high ($W_V \geq 10^{-1}$ J/cm^3) for electromagnetic rotational motors, pneumatic linear and rotational actuators, piezoelectric inchworm actuators and shape memory actuators.

However, work density per cycle is low ($W_V \leq 10^{-3}$ J/cm^3) for piezoelectric multimorph actuators, magnetostrictive actuators and travelling wave linear ultrasonic motors (TWLUM). It is moderate ($10^{-1} \geq W_V \geq 10^{-3}$ J/cm^3) for the remaining technologies, that is, piezoelectric stack actuators, electromagnetic linear motors, travelling wave rotational ultrasonic motors (TWRUM) and magnetorheological fluid actuators.

7.2.5 Comparative analysis in terms of power density

Power density has been defined as the ratio of maximum output power (dissipative power for semiactive actuators) to the volume or weight of the actuator. It is analyzed in this section for all the actuator technologies described in this book.

Power density is high ($P_V \geq 10$ W/cm^3) only for piezoelectric stacked actuators. It may be considered moderate ($10 \geq P_V \geq 10^{-1}$ W/cm^3) for piezoelectric multimorph actuators, pneumatic linear and rotational actuators, electromagnetic linear and rotational motors, travelling wave linear and rotational ultrasonic motors (TWLUM and TWRUM), magnetostrictive actuators and magnetorheological fluid actuators. It is low ($P_V \leq 10^{-1}$ W/cm^3) for SMA actuators and piezoelectric inchworm actuators.

According to the scaling analysis for most technologies, the power density scales between $P_V \propto L^{-1}$ and $P_V \propto L^{-2}$. Figure 7.6 shows trend lines for this theoretical result to give an idea of the accuracy of this prediction. The experimental data fit the theoretical result well enough, although the number of specimens for each technology should be higher for conclusive results.

7.2.6 Comparative analysis in terms of bandwidth

This section presents a comparison of the different emerging actuators in terms of the maximum actuation frequency they can withstand. The theoretical scaling laws for the actuator's bandwidth vary between $f \propto L^{-2}$ and $f \propto L^{-1}$. The corresponding trend lines are indicated in Figure 7.7. In general, the smaller the actuator, the higher is the frequency bandwidth.

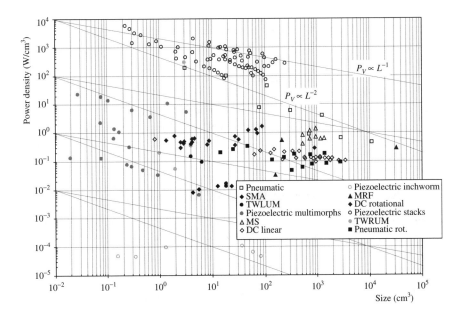

Figure 7.6   Comparison of emerging and traditional actuators in terms of power density.

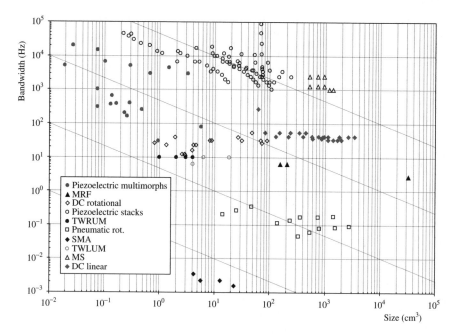

Figure 7.7   Comparison of emerging and traditional actuators in terms of bandwidth.

The chart in Figure 7.7 shows bandwidth versus size for all emerging actuator technologies, and also for electromagnetic and pneumatic linear and rotational drives.

Piezoelectric stack and multimorph actuators and magnetostrictive actuators may be considered fast technologies ($f \geq 10^2$ Hz). At the opposite extreme, pneumatic rotative actuators and SMAs are slow actuators ($f \leq 1$ Hz). All the other technologies, that is, linear and rotational electromagnetic motors, travelling wave linear and rotational ultrasonic motors and magnetorheological actuators have moderate bandwidth ($1 \leq f \leq 10^2$ Hz).

7.2.7   Relative position in the static and dynamic plane

The graphs appearing in all the previous sections can be used to compare emerging and traditional actuator technologies and are suitable for evaluating the following:

1. *The relative position for a single figure of merit.*

2. *The trend of each actuator technology for a single figure of merit upon scaling.*

This section introduces a comprehensive global comparison of emerging and traditional actuator technologies. It is based on the following:

1. *The static-plane representation.* The static plane is defined as the graph representing maximum stress versus relative stroke for all actuator technologies. Wherever experimental data are not available, theoretical upper-bound projections are included.

2. *The dynamic-plane representation.* This is the graphical representation of energy density per cycle versus power density for all actuators. Again, theoretical projections are used in the case of new technologies for which experimental data are not available.

The static-plane representation for all actuator technologies is depicted in Figure 7.8. In addition to all the emerging actuators discussed in Sections 7.2.1 to 7.2.3, theoretical projections for conducting polymer actuators, dielectric elastomer actuators and polymer gel actuators have been included.

Since EAPs are commonly referred to as artificial muscles, Figure 7.8 also includes the operational ranges for human muscle.

The static plane is helpful for classifying actuators into four categories:

1. *High-force, low-stroke actuators.* This operational range corresponds to the top-left quadrant in the static plane. This region includes mainly piezoelectric stack actuators.

2. *High-force, high-stroke actuators.* This operational range corresponds to the top-right quadrant of the static plane. Shape memory alloys fall within the

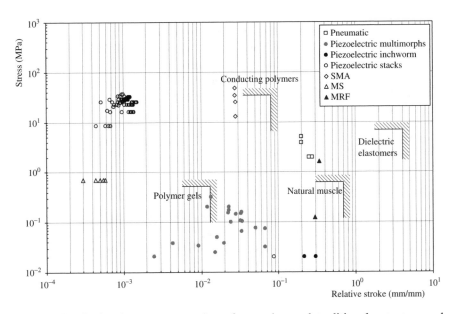

Figure 7.8   Static-plane representation of emerging and traditional actuators and theoretical prediction for EAP actuators.

area limiting this and the previous region. Also, pneumatic actuators belong in the area limiting this and the next region (low force, high stroke). Theoretical predictions for CP and dielectric elastomer actuators are also in this region.

3. *Low-force, high-stroke actuators.* This operational range corresponds to the bottom-right quadrant of the static plane. Piezoelectric inchworm actuators and magnetorheological actuators belong in this operational range, as do human muscle characteristics also.

4. *Low-force, low-stroke actuators.* This operational range corresponds to the bottom-left quadrant of the static plane. It includes magnetostrictive actuators, piezoelectric multimorph actuators and the theoretical projection for polymer gel actuators.

It is worth noting that the theoretical predictions for EAP actuators can readily be reduced by several orders of magnitude when practically implemented. These projections are based on the transducing material itself without taking any accompanying component into account.

To illustrate this effect, let us recall the case of magnetostrictive actuators. If the transducing material is considered independently, its stroke should be larger than that of piezoelectric stack actuators. If we look at Figure 7.8, it is apparent that this is not so in a practical implementation of the technology. The actual stroke has been reduced by one order of magnitude in the process of practical implementation (in which all accompanying elements are included, for example, prestress mechanism, magnetic field coil, permanent magnets for bias field...).

Figure 7.9 shows the dynamic-plane representation for all the various emerging and traditional actuator technologies. The theoretical prediction for most EAP actuators and for human muscle is also included. Like the static-plane representation, the dynamic plane is useful for classifying the different actuator technologies in terms of their dynamic operational range:

1. *High work density, low power density actuators.* This operational range corresponds to the top-left quadrant in the dynamic plane. It roughly covers slow actuators delivering high force and comprises SMA actuators, pneumatic rotational actuators and electromagnetic rotational actuators.

2. *High work density, high power density actuators.* This operational range corresponds to the top-right quadrant of the dynamic plane. The region includes fast drives delivering moderate to high force. Pneumatic linear actuators, in particular, lie at the limit between the previous and this region, and most of the theoretical predictions for EAP actuators fall within this operational region.

3. *Low work density, high power density actuators.* This operational range corresponds to the bottom-right quadrant of the dynamic plane. Piezoelectric

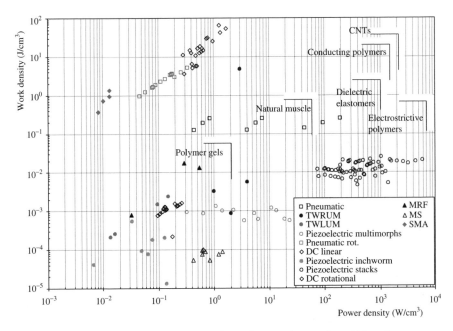

Figure 7.9 Dynamic-plane representation of emerging and traditional actuators, and theoretical prediction for EAP actuators.

stack actuators can be found at the upper limit of this region, while piezo-electric multimorph actuators lie at the rightmost limit, effectively entering the low power density region.

4. *Low work density, low power density actuators.* This operational range corresponds to the bottom-left quadrant of the dynamic plane. Travelling wave linear ultrasonic motor, electromagnetic linear motor and magnetorheological fluid actuators are included in this operational range. Magnetostrictive actuators and travelling wave rotational actuators lie at the boundary with the previous operational region.

7.2.8 Comparison in terms of scaling trends

The scaling trends for each emerging actuator technology were analyzed in detail in the relevant chapters. Table 7.1 summarizes the scaling laws for all technologies.

All the actuator technologies described in this book share the same scaling laws for force, $F \propto L^2$, force density, $F_V \propto L^{-1}$, stroke, $S \propto L$, energy per cycle, $W \propto L^3$, and energy density per cycle, $W_V \propto L^0$. This means that for all the actuator technologies:

- The force increases quadratically when the actuator's dimensions are scaled up.

Table 7.1   Summary of scaling trend for emerging actuator technologies.

Technology	F	F_V	S	W	W_V	f	P_V
Piezoelectric	L^2	L^{-1}	L	L^3	L^0	L^{-1}	L^{-1}
SMA	L^2	L^{-1}	L	L^3	L^0	L^{-2}	L^{-2}
EAP (wet)	L^2	L^{-1}	L	L^3	L^0	L^{-2}	L^{-2}
EAP (dry)	L^2	L^{-1}	L	L^3	L^0	L^{-1}	L^{-1}
Magnetostrictive	L^2	L^{-1}	L	L^3	L^0	L^{-1}	L^{-1}
MRF–ERF, F_τ (F_η)	L^2 (L)	L^{-1} (L^{-2})	L	L^3	L^0	L^{-1}	L^{-1}

- The force density increases as the dimensions are reduced.

- The stroke increases linearly with increasing dimensions.

- The energy per cycle increases cubically when the dimensions are scaled linearly.

- The energy density per cycle remains roughly constant when the actuator is scaled either up or down.

For the particular case of ERF and MRF, the viscous force varies linearly with dimensions, $F_\eta \propto L$. The field-dependent force, F_τ, follows the same law as all other technologies.

Where actuator technologies differ most is in their dynamic properties. For all actuator technologies, whose time response is limited by transport mechanisms (i.e. SMA and wet EAPs), the bandwidth scales as $f \propto L^{-2}$. For all other actuators, the bandwidth scales as $f \propto L^{-1}$. This indicates that the impact of miniaturization in improving time response is comparatively larger for mass or thermal transport – limited actuators than for other technologies. The power density scales like the bandwidth.

The scaling trends of the different actuator technologies can be seen in charts 7.2 to 7.6. These charts indicate the theoretical trends so as to simplify interpretation of the results. Graphs 7.2 to 7.5 show only one trend line (since it is common to all actuator technologies), while graphs 7.6 and 7.7 show both inverse linear and inverse quadratic trends.

7.2.9 Concluding remarks

This comparison has addressed the various figures of merit introduced in Chapter 1 and discussed throughout the different chapters, namely, force, force density, torque, stroke, work density per cycle, power density and bandwidth. In each graph, trend lines have been plotted according to the scaling analysis so that the accuracy of predictions based on scaling laws can be verified.

The comparative analysis yields the following conclusions:

1. *Emerging actuators complement traditional ones.* There is no significant overlapping of performance (either static or dynamic). Emerging and traditional actuators should be seen as complementary rather than competing devices.

2. *Emerging actuators span the static and dynamic planes.* Emerging actuators complement one another and fill the gaps left by traditional technologies both in static and dynamic characteristics.

3. *The scaling laws are roughly verified.* The scaling trends for the diverse actuator technologies and figures of merit have been verified where enough data points are available. For those actuators with insufficient data points, the trends cannot be considered verified.

4. *Theoretical predictions greatly overestimate practical results.* In the case of MS actuators, it has been shown that actual implementations result in figures of merit that are up to several orders of magnitude lower than theoretical projections. This is particularly relevant in the case of EAP actuators, for which only scanty experimental data are available.

7.3   Research trends and application trends

In previous sections and chapters, we have analyzed the state of the art in emerging actuator technologies. The book now concludes with an overview of research and application trends in the different emerging technologies.

A complete and detailed list of current or future research trends in the various different actuator technologies is beyond the scope of this book. Here, we will confine ourselves to highlighting those aspects that in our opinion merit further study with a view to making actuators robust enough for use in new fields of application.

When looking at research trends and application domains, it is worth noting that the way emerging actuator technologies are applied will be radically different from the approach adopted with traditional actuators. Traditional actuators are commercial, off-the-shelf components. When configuring motion control systems, engineers can choose from among *de facto* standardized actuators, transmission stages and driving electronics.

The approach in emerging actuator technologies is mostly quite the opposite. The motion control system (including the actuator, the impedance-matching subsystems and the control) is engineered to meet the application characteristics. This results in a more nuanced approach as it lends additional flexibility to the process of optimizing one particular solution; however, it means that the engineer must have a thorough grounding in mechatronics.

Some attempts have been made to mimic the design process of traditional actuator technologies with emerging actuator technologies. Examples of this second approach include piezoelectric actuators and magnetorheological fluid actuators.

7.3.1 Piezoelectric actuators

Piezoelectric actuators are the most mature technology of the various different emerging actuators, in terms both of the material itself and of the actuator configurations and concepts. Following are some lines of research that we may expect to see in the near future:

1. *Material research.* Even though piezoelectric materials have been studied for decades, some issues still remain. One goal that researchers particularly keep in mind is to increase the transduction level (increase resulting stroke) by means of single-crystal piezoelectric ceramics.

   Another trend in research is the combination of additional transducing processes in piezoelectric materials. In particular, the inclusion of magneto-piezoelectricity and photostriction would increase the possibility of integrating sensing and actuation structures.

   And, likely research lines in the improvement of material characteristics are the pursuit of high-temperature piezoelectric materials (with Curie temperature in excess of $500\,^\circ$C) and reinforced ceramics through grain boundary engineering.

2. *Composite piezoelectric materials.* In the past, a great deal of effort has been devoted to improving piezoelectric transducers by means of composite structures (Moonie and Cymbal composite transducers). The pursuit of new composite approaches will probably be a hot research area in the future.

3. *New actuation concepts.* The application of new actuation configurations, in particular, for resonant piezoelectric actuators is an open field of research. This includes the efficient combination of different resonant modes in which to render actuation mechanisms.

The application domain for piezoelectric actuators is as wide as the variety of actuator concepts within the broad area of this technology. *Precision positioning* (in the context of optics or microsystem technologies), for instance, is a classic field of application for piezoelectric stack actuators, as they can supply sub-nanometer resolution with an acceptable stroke.

In general, all emerging actuator technologies are appropriate for developing the concept of *smart actuators and smart structures*, but this is particularly true of piezoelectric actuators, as they can be integrated in passive structures to make true functional structures. By developing the concept of concomitant sensing and actuation (i.e. collocated sensors and actuators), the control of these structures (in active vibration suppression schemes) can be rendered intrinsically stable.

Low amplitude, high power *active vibration suppression* is another typical application domain. This approach can be implemented in a number of situations, for instance, in high-precision machine finishing processes.

7.3.2 Shape memory alloy actuators

SMA actuators are beginning to be implemented in industrial applications. Their status in terms of material development and actuator configurations is well established, but there is still room for research. Most of the work is addressing the pursuit of new applications. Following are some of the research topics:

1. *Material research.* One of the limiting factors in SMA actuators is the relatively low fatigue limits of these materials. A promising line of research for the future is the improvement of material properties in terms of fatigue by means of either training processes or ternary additions to the basic NiTi alloy.

   Another important item on the agenda for the future is the engineering of material characteristics to suit the application. This is closely related to the issue of precise control of material thermal properties (in particular, transformation temperatures and their relative position). This is expected to have a considerable impact on the applicability of SMA actuators in the automotive industry.

   In applications where the pseudoelastic properties of SMAs are exploited, we may expect to see intense research work on single-crystal materials in a drive to increase recoverable strains.

2. *New actuation mechanisms.* The boundary conditions being imposed on the SMA actuator may have a considerable impact on the overall performance of the actuator. New concepts for the application of bias force to actuators will be analyzed in the coming years. We may also expect to see new actuation mechanisms in addition to the classic tensile, bending or torsion concepts.

We foresee a search for new applications for active devices that will exploit the self-sensing capabilities of SMA actuators, their intrinsically high force density and their compact configurations. However, we believe that the most appropriate application domains are the following:

1. *Passive thermal actuators.* SMA actuators are particularly suited to passive actuation in response to changes in ambient temperature; we see this type of application as the best suited to the technology. A typical application following this approach was introduced in Case Study 3.2.

   Passive thermal actuators directly exploit the transduction phenomenon in SMAs. Efficiency is not a concern in this type of applications; moreover, they generally require slow actuation mechanisms. All these application characteristics match the actuation conditions of SMA actuators.

2. *One-time actuation applications.* Fatigue behavior of SMAs is a serious problem in active actuator applications. Until significant improvements are achieved in the fatigue life of these alloys, single actuation applications (or at least applications where the number of cycles is limited; see Case Study 3.1) will continue to be the most appropriate.

3. *Smart structures.* This is a promising field given the combination of material characteristics that can be exploited in the context of smart structures (e.g. change in Young's modulus, superelasticity and concomitant sensing and actuation).

7.3.3   Electroactive polymer actuators

EAP actuators embrace a multiplicity of dissimilar technologies, which are also at different stages of development. Dry EAPs are better established than wet EAP technologies. Following is a summary of the areas requiring further study:

1. *Understanding the transduction process.* EAP actuators usually exhibit complex nonlinear, hysteretic and discontinuous behavior, which is not well understood. Understanding the actuation process and how the various different material and design parameters influence the actuator's performance and characteristics is a prerequisite for an application-oriented development of EAP actuators.

   Within this field, accurate *models of the transduction process* are required. This will probably require multidimensional models covering electro-mechano-chemical interactions in both static and dynamic operations.

2. *Mastering and engineering materials.* This requires a good understanding of synthesis, properties of materials and actuator performance. New means of *processing and actuator manufacturing* are required.

   One of the limiting factors in ionic EAPs is their intrinsically slow response, which is limited by ion diffusion processes. Special attention will probably be required in the future to *improve ion mobility in the polymer structure.*

   Another research area is *new materials for electrodes.* There is an intrinsic mismatch between strain characteristics of EAPs and those of the metallic electrodes that are used. This results in limited lifetimes owing to delamination. Further study and synthesis of new electrodes are required.

3. *Technologies for actuator development.* The engineering of actuators from EAP materials is highly complex. *New geometrical transducing mechanisms* will probably emerge in the near future. If research addresses the application of ionic EAP actuators in dry environments, *packaging will become a crucial issue.*

   Control technologies for these new actuators will have to be analyzed and proposed, particularly to deal with hysteresis, relaxation processes, dissimilar dynamics upon contraction and relaxation, and discontinuous operation due to phase transitions (chiefly in polymer gel actuators).

Appropriate impedance-matching mechanisms will be required in the electrical and mechanical ports. If EAP actuators are to be implemented in applications not exactly matching their actuation characteristics, motion transmission mechanisms will have to be developed.

As regards foreseeable application fields, EAP actuators are expected to perform properly under conditions matching their actuation characteristics. The following is a review of some tentative application domains for EAP actuators:

1. *Biomedical applications.* As we discussed in Case Study 4.1, CP actuators are especially suitable for biomedical applications in which the actuators have to be in contact with body fluids. CPs are biocompatible materials both *in vitro* and *in vivo*. They have proved to be tissue compatible for long periods and in some instances can be made biodegradable.

   CP actuators are particularly suitable for low-force, high-stroke, one-time actuation; that is the actuation condition in most biomedical applications. Other EAP technologies are not particularly suited to this type of application: dry EAP technologies require high voltages for operation, while other wet EAP technologies (particularly polymer gel actuators) do not meet biocompatibility requirements.

2. *Vibration isolation.* The passive counterparts of some EAP materials are currently used as passive vibration isolation components. One such case is dielectric elastomers, which suggests that dielectric elastomer actuators could be used in active vibration isolation.

3. *Other application fields.* Some EAP actuator technologies are so new that specific application fields have yet to be identified. In these circumstances, the logical response is to propose them as competitor technologies for either traditional actuators or other emerging actuators. They have therefore been proposed for use in robotics, prosthetics (because their behavior is similar to that of human muscle), animatronics and so forth.

   This approach has specific drawbacks, particularly in the case of wet EAP actuators. For these actuators, the current technological difficulties (packaging, material engineering, etc.) pose too serious a challenge. In our opinion, these new technologies must target intrinsically matched environments (e.g. ionic EAP for saline environments, such as in underwater applications).

7.3.4 Magnetostrictive actuators

Magnetostriction has been known since the nineteenth century, but new magnetostrictive materials are currently an active area of research. With the advent of twin induced magnetostriction as a new transduction phenomenon, there has been more emphasis on materials research. The following are some of the research fields in which we may expect to see most action in the coming years:

1. *Development of new magnetostrictive materials.* It is likely that research activities will focus on developing new materials for operation in harsh and demanding environments as well as materials with lower requirements in terms of magnetic field strength. This applies, for instance, to new magnetostrictive materials like Galfenol. One of the difficulties that the scientific community will need to overcome is to render lower magnetic field requirements compatible with similar stroke levels.

2. *New twin-based magnetostrictive materials.* The transduction process in twin induced magnetostriction is closely related to the material's magnetic anisotropy. New heat treatments and the addition of ternary and quaternary elements to the basic alloy composition in these materials seem likely research fields.

3. *Improved operational characteristics.* Research on operational characteristics of magnetostrictive actuators must be paralleled by improvements in the material's characteristics and in control and application aspects. As regards the material, higher actuation bandwidth will be possible if eddy currents are limited by decreasing the material's conductivity.

4. *Improved control.* In order to fully exploit the actuation features of MS actuators, they need to be better integrated in control loops. This is true, for instance, of improved utilization of the self-sensing capabilities of MS actuators, their integration in collocated sensor actuator approaches and in the pursuit of new applications that exploit other aspects of MS actuators such as the change in Young's modulus, resonance frequency and other allied phenomena.

The field of application for MS actuators has hitherto been largely high-power active vibration control in aviation, civil engineering and structures, and this will most likely continue to be the main domain for the application of MS actuators. However, new applications are emerging and may become alternatives in due course. This is the case in the automotive industry, where novel applications in active control of fuel injection have been proposed and will certainly be the subject of further research.

7.3.5 Electro- and Magnetorheological fluid actuators

Chapter 6 dealt with semiactive actuators based on electro- and magnetorheological fluid actuators. As noted in Chapter 6, ER fluid actuators have a number of drawbacks with respect to their MR counterparts. In particular, ERF actuators present low yield stress, require strong electric fields, are chemically unstable and possess highly temperature-dependent electrical properties (in particular, conductivity). As a result, they present serious problems for practical application.

MRF actuators, on the other hand, are well suited for practical applications and have already been implemented industrially in various sectors. Given the present status of ERF and MRF actuators, considerable research is still required in the area of ERF actuators (at all levels), while in the case of MRF actuators, we are likely to see a search for new applications. The following are some probable research trends:

1. *Increased yield strength ER fluids.* One of the most serious impediments to the practical applicability of ERF actuators is their low yield strength (two orders of magnitude less than MR fluids). Recent research on urea nanocoated barium titanyl oxalate particles (Wen *et al.* (2003)) has brought the yield strength of ER fluids (known as giant electrorheological fluids, GER fluids) up close to that of MR fluids. Research in this line is expected to intensify, focusing on fluids that may be able to overcome the chief limiting factors in the applicability of ER fluid actuators.

2. *Chemical stability and dependence on external factors.* For practical purposes, ER fluids must be chemically stable and their dependence on external factors (chiefly temperature) must be limited. We may expect intense research activity in this area, in the coming years, aimed at overcoming these problems.

3. *Aspects related to the practical implementation.* If strong electric fields are still required to drive ER fluid actuators, this will pose serious practical problems. For instance, strong electric fields can seriously compromise wiring and connector solutions for ER fluid actuators. Also related to the practical use of ER fluid actuators is the need for electrical driving circuits and control strategies to overcome the changing electrical properties of ER fluid actuators in response to external perturbations (changes in temperature of the fluid).

4. *Mathematical models.* At a more theoretical level, although with clear practical implications, there is a need to develop new, accurate dynamic models of ER and MR fluid actuators and of the electro- and magnetorheological effect. This will improve insight and understanding, which in turn will improve practical application (actuators better adapted to the application and control strategies better adapted to actuator characteristics).

As to the field of application of ER and MR fluid actuators, as semiactive actuators, this will be mostly limited to semiactive vibration control and to devices in which damping and stiffness properties can be controlled in response to requirements. The first application field has already been explored in the case of MR actuators and several applications in the automotive industry are ongoing. If the above research activities produce practical fluids, it will be time to explore the field for ER fluid actuators.

The second application domain, in which the ER or MR fluid is integrated in a responsive structure (change in damping and stiffness) is being explored in the field of rehabilitation (see the lower-limb prosthetic solution in Case Study 6.1 and Case Study 6.2). New applications developing this concept are expected to appear in the coming years.

Bibliography

Abadie J, Chaillet N, Lexcellent C and Bourjault A 1999 Thermoelectric control of shape memory alloy microactuators: a thermal model, *Proceedings of SPIE*, **3667**(1), 326–336.

Bartlett PA, Eaton SJ, Gore J, Matheringham WJ and Jenner AG 2001 High-power, low frequency magnetostrictive actuation for anti-vibration applications, *Sensors and Actuators A*, **91**, 133–136.

Baughman RH, Cui C, Zakhidov AA, Iqbal Z, Barisci JN, Spinks GM, Wallace GG, Mazzoldi A, De Rossi D, Rinzler AG, Jaschinski O, Roth S and Kertesz M 1999 Carbon nanotube actuators, *Science*, **284**, 1340–1344.

Bay L, Jacobsen T, Skaarup S and West K 2001 Mechanism of actuation in conducting polymers: osmotic expansion, *Journal of Physical Chemistry B*, **105**, 8492–8497.

de Boer E 1961 Theory of motional feedback, *IRE Transactions on Audio*, **AU-9**, 15–21.

Brenner W, Mitic S, Vujanic A and Popovic G 2000 Micro-actuation principles for high resolution graphic tactile displays, *Actuator 2000*, Bremen, 567–570.

Busch-Vishniac IJ 1998 *Electromechanical Sensors and Actuators*, Springer, Berlin.

Claeyssen F, Lhermet N and Maillard T 2002 Magnetostrictive actuators compared to Piezoelectric actuators, *ASSET 2002*, http://www.cedrat.com, last accessed September 2004.

Clark AE and Savage HT 1975 Giant magnetically induced changes in the elastic moduli in Tb(0.3)Dy(0.7)Fe(2), *IEEE Transactions on Sonics and Ultrasonics*, **SU-22**(1), 50–52.

Dapino MJ, Smith RC, Calkins FT and Flatau A 2002 A coupled magnetomechanical model for magnetostrictive transducers and its application to Villari-effect sensors, *Journal of Intelligent Material Systems and Structures*, **13**, 737–747.

Dörlemann C, Muss P, Schugt M and Uhlenbrock R 2002 New high speed current controlled amplifier for PZT multilayer stack actuators, *Actuator 2002*, Bremen, 337–340.

Dosch JJ, Inman DJ and García E 1992 A self-sensing piezoelectric actuator for collocated control, *Journal of Intelligent Material Systems and Structures*, **3**, 166–185.

Duering TW, Melton KN, Stöckel D and Wayman CM 1990 *Engineering Aspects of Shape Memory Alloys*, Butterworth-Heinemann, London.

Fenn RC, Downer JR, Bushko DA, Gondhalekar V and Ham ND 1996 Terfenol–D driven flaps for helicopter vibration reduction, *Smart Materials and Structures*, **5**, 49–57.

Flatau AB, Dapino MJ and Calkins FT 1998 High bandwidth tunability in a smart vibration absorber, *SPIE Smart Structures and Materials Conference*, San Diego, 463–473.

Gardner FM 1979 *Phaselock Techniques*, John Wiley & Sons, New York.

Emerging Actuator Technologies: A Micromechatronic Approach   J. L. Pons
© 2005 John Wiley & Sons, Ltd

Genç S 2002 Synthesis and Properties of Magnetorheological (MR) Fluids, Ph.D. Thesis, University of Pittsburgh.

Graff KF 1975 *Wave Motion in Elastic Solids*, Dover Publications, New York.

Grant D and Hayward V 1997 Variable structure control of shape memory alloy actuators, *IEEE Control Systems Magazine*, **17**, 80–88.

Hagwood, NW and Anderson EH 1991 Simultaneous sensing and actuation using piezo-electric materials, *Active and Adaptive Optical Components, SPIE*, **1543**, 409–421.

Harrison JD 1990 Measurable changes concomitant with the shape memory effect transformation, 106–111, in *Engineering Aspects of Shape Memory Alloys*, Duering TW, Melton KN, Stöckel D and Wayman CM Eds, Butterworth-Heinemann, London.

Hasegawa T and Majima S 1998 A control system to compensate the hysteresis by Preisach model of SMA actuator, *Proceedings of the 1998 9th International Symposium on Micromechatronics and Human Science*, 171–176.

de Heer WA 2004 Nanotubes and the pursuit of applications, *MRS Bulletin*, **29**(4), 281–285.

Hesselbach J, Pitshellis R and Stork H 1994 Optimization and control of electrically heated shape memory actuators, *Actuator '94*, Bremen, 337–340.

Hogan N 1985a Impedance control: an approach to manipulation (Part I – Theory), *Journal of Dynamic Systems, Measurement and Control*, **107**, 1–7.

Hogan N 1985b Impedance control: an approach to manipulation (Part II – Implementation), *Journal of Dynamic Systems, Measurement and Control*, **107**, 8–16.

Hunt FV 1982 *Electroacoustics: The Analysis of Transduction and its Historical Perspective*, Acoustical Society of America.

Ikuta K, Tsukamoto M and Hirose S 1988 Shape memory alloy servo actuator system with electric resistance feedback, *Proceedings of the IEEE International Conference on Robotics and Automation*, Computer Society Press, Washington, DC, 427–430.

Immerstrand C, Holmgren-Peterson K, Magnusson KE, Jager E, Krogh M, Skoglund M, Selbing A and Inganä O 2002 Conjugated-Polymer micro- and milliactuators for biological applications, *MRS Bulletin*, **27**(6), 461–464.

Infinite Biomedical Technologies, Baltimore, http://www.i-biomed.com/index.html, last accessed September 2004.

Intelligent Polymer Research Institute, Wollongong, http://www.uow.edu.au/science/research/ipri/, last accessed September 2004.

Janocha H 2001 Application potential of magnetic field driven new actuators, *Sensors and Actuators A*, **91**, 126–132.

Jolly MR and Carlson JD 1996 Controllable squeeze film damping using magnetorheological fluid, *Actuator 1996*, Bremen, 333–336.

Karnopp D, Crosby M and Harwood RA 1974 Vibration control using semi-active suspension control, *Journal of Engineering for Industry*, 619–626.

Kakeshita T and Ullakko K 2002 Giant magnetostriction in ferromagnetic shape-memory alloys, *MRS Bulletin*, **27**, 105–109.

Kuhnen K and Janocha H 1998 Compensation of the creep and hysteresis effects of piezo-electric actuators with inverse systems, *Actuator '98*, Bremen, 309–312.

Lee HJ and Lee JJ 2004 Time delay control of shape memory alloy actuator, *Smart Mater. Struct.*, **13**, 227–239.

Landau LD and Lifshitz EM 1970 *Electrodynamics of Continuous Media*, Pergamon Press, Oxford.

Lord Corporation 1999 Designing with MR fluids, Engineering Note, November 1999, http://www.lord.com.

Ma N and Song G 2003 Control of shape memory actuator using pulse width modulation, *Smart Materials and Structures*, **12**, 712–719.

Madden J 2004 Properties of electroactive polymer actuators, *Actuator 2004*, Bremen, 338–343.

Madou M 1997 *Fundamentals of Microfabrication*, CRC Press.

Mertmann M, and Wuttig M 2004 Application and hysteresis of different shape memory alloys for actuators, *Actuator 2004*, Bremen, 72–77.

Mertmann M, Kautz S and Brown W 2002 Design principles and new actuator applications with NiTi straight wire actuators, *Actuator 2002*, Bremen, 85–90.

Middlehoek S and Hoogerwerf AC 1985 Smart sensors: When and Where? *Sensors and Actuators*, **8**(1), 39–48.

Mitwalli AH 1998 Polymer gel actuators and sensors, Ph.D. Thesis, Massachusetts Institute of Technology.

Nemat-Nasser S and Thomas CW 2001 Ionomeric polymer-metal composites, 139–191, in *Electroactive Polymer (EAP) Actuators as Artificial Muscles*, Bar-Cohen Ed., SPIE Press, Washington, DC.

Otsuka K and Kakeshita T 2002 Science and technology of shape-memory alloys: new developments, *MRS Bulletin*, **27**(2), 91–100.

Peirs J 2001 Design of micromechatronic systems: scale laws, technologies, and medical applications, Ph.D. Thesis, Katholieke Universiteit, Leuven.

Pérez-Aparicio JL and Sosa H 2004 A continuum three-dimensional, fully coupled, dynamic, non-linear finite element formulation for magnetostrictive materials, *Smart Materials and Structures*, **13**, 493–502.

Phillips RW 1969 Engineering applications of fluids with variable yield stress, Ph.D. Thesis, University of California, Berkeley.

Pratt J 1993 Design and analysis of a self-sensing Terfenol–D magnetostrictive actuator, M.Sc. Thesis, Iowa State University, Ames.

Preumont A 1997 *Vibration Control of Active Structures*, Kluwer Academic Publishers.

Preumont A and De Man P 2004 Some aspects of active and semi-active vibration isolation, *Actuator 2004*, Bremen, 1–6.

Rabinow J 1948 The magnetic fluid clutch, *AIEE Transactions*, **67**, 1308–1315.

Reynaerts D 1995 Control methods and actuation technology for whole-hand dexterous manipulation, Ph.D. Thesis, Katholieke Universiteit, Leuven.

Reynaerts D, Peirs J and Van Brussel H 1998 A mechatronic approach to microsystem design, *IEEE/ASME Transactions on Mechatronics*, **3**(1), 24–33.

Rosen CZ, Hiremath BV and Newnham R, Eds 1992 *Piezoelectricity*, American Institute of Physics, New York.

Rosenberg RC and Karnopp DC 1983 *Introduction to Physical System Dynamics*, McGraw-Hill, New York.

Sansiñena JM and Olazábal V 2001 *Conductive polymers*, 193–221, in *Electroactive Polymer (EAP) Actuators as Artificial Muscles*, Bar-Cohen Ed., SPIE Press, Washington, DC.

Schaaf U and van der Broeck H 1995 Piezoelectric motor fed by a PLL-controlled series resonant converter. In *Proceedings of ENE '95*, Sevilla, 3845–3850.

Smela E 2003 Conjugated polymer actuators for biomedical applications, *Advanced Materials*, **15**(6), 481–494.

Smela E and Gadegaard N 2001 Conjugated Polymer actuators for biomedical applications, *Journal of Physical Chemistry B*, **105**, 9395.

Sommer-Larsen P and Kornbluh R 2004 Polymer actuators, *Actuator 2004*, Bremen, 371–378.

Song G, Chaudhry V and Batur C 2003 Precision tracking control of shape memory alloy actuators using neural networks and sliding-mode based robust controller, *Smart Materials and Structures*, **12**, 223–231.

Stiebel C and Janocha H 2000 6 kV Power amplifier designed for actuators with electrorheological (ER) fluids, *Actuator 2000*, Bremen, 135–138.

Thrasher MA, Shalin AR and Meckl PH 1994 Efficiency analysis of shape memory actuators, *Smart Materials and Structures*, **3**, 226–234.

Uchino K and Hirose S 2001 Loss mechanisms in piezoelectrics: how to measure different losses separately, *IEEE Transactions on Ultrasonics, Ferroelectrics, and Frequency Control*, **48**(1), 307–321.

Ueha S, Tomikawa Y, Kurosawa M and Nakamura N 1993 *Ultrasonic Motors. Theory and Applications*, Clarendon Press, Oxford.

Wallace GG, Ding J, Lu L, Spinks GM, Zhou D, Forsyth S and Forsyth M 2002, *SPIE, Smart Structures and Devices, EAPAD*, 4695.

Wax SG and Sands RR 1999 Electroactive polymer actuators and devices, *SPIE Conference on Electroactive Polymer Actuators and Devices*, 2–10.

Wen W, Huang X, Shihe Y, Lu K and Sheng P 2003 The giant electrorheological effect in suspension of nanoparticles, *Nature Materials*, **2**, 727.

Winslow WM 1949 Induced fibration of suspensions, *Journal of Applied Physics*, **20**(12), 1137–1140.

Yang G 2001 Large-scale magnetorheological fluid damper for vibration mitigation: modeling, testing and control, Ph.D. Thesis, Department of Civil Engineering and Geological Sciences, Notre Dame.

Index